RATS

It was a nice pink pile carpet. . . The door to the baby's nursery room was open and the baby lay crying in its cot. Around its mouth were traces of milk – the baby had just been fed and had regurgitated some of its feed. The rat had climbed into the cot and started licking the milk off the baby's face. The baby moved its head in the wrong direction, and the rat bit. . .

Mark Jervis
Pest Control Supervisor,
London Borough of Barnet

RATS
THE NEW PLAGUE

Charles Golding

Weidenfeld and Nicolson
London

PICTURE ACKNOWLEDGEMENTS

The publishers are grateful to the following for permission
to reproduce the photographs:
Aquila Photographics 2, 10; NHPA 1, 5, 6, 14; Rentokil
3, 8, 9, 12, 13; Rex Features 7; Survival Anglia 4, and
Jackie Hyman 15.

Picture research by Ivor Game

Published in Great Britain by
George Weidenfeld & Nicolson Limited
91 Clapham High Street
London SW4 7TA

ISBN 0 297 81082 0

Photoset by Deltatype Ltd, Ellesmere Port, Cheshire
Printed in Great Britain by
The Guernsey Press Co. Ltd, Guernsey, C.I.

Contents

This book is dedicated to my mother and father whose constant enthusiasm and practical support helped me make it into a reality and them into reluctant experts in rodent affairs.

Acknowledgements

I would like to express my gratitude to all of the many people who have assisted me in writing this book. First and foremost, I would like to thank my enthusiastic research assistant, Jackie Hyman. Particular mention must also be made of Dr Sheena Waitkins, Peter Bateman and Bob Tanner, all of whom worked with me beyond the call of duty, and Dr Ian Hatrick of Guildford, Surrey, for his advice on Weil's disease.

I would also like to thank several institutions and companies that helped me in my research; the Leptospira Reference Unit; the Institution of Environmental Health Officers; the London Borough of Barnet; Manchester City Council; the Museum of London; the British Canoe Union; Rentokil Plc and Regent Publishing.

Last but not least, my special thanks go to David Drake, Rob Mylchreest, Harry Beasley, Mark Jervis and all of the rodent control operators for taking me down the drains and into the sewers, showing me that the art of ratcatching is still alive and well in Britain today.

Introduction

Rats have always held a morbid fascination for mankind. They are almost universally loathed and feared, not without good reason. Rats still carry disease and destruction, just as they carried the plague-infected flea, *Xenopsylla cheopsis*, into fourteenth century Europe bringing death and misery on a scale previously unknown. The Black Death killed about 1.2 million people – a third of the population of medieval England – and *Rattus rattus*, the black rat, was largely responsible for its transmission. Black rats no longer inhabit Britain. They were usurped by a larger, fiercer breed known as *Rattus norvegicus*, the brown rat. Born to survive, brown rats took less than ten years to infest our shores in the eighteenth century, driving out their black cousins and establishing a permanent foothold. They continue to rise in number today and little is being done to stop them.

It has been almost impossible to pick up a newspaper over the last twelve months without venturing across another story about rats. Millions of rats, we read, are surfacing from sewers all over the United Kingdom, from pig sties to palaces, at home and at work, in the town and in the country. The incidence of rat bites and salmonellosis directly traced to rats is on the increase. Most frightening of all, cases of leptospirosis – Weil's disease – have risen significantly this year, resulting in fifteen known deaths and hundreds more seriously ill victims.

Rats are known as commensal animals, meaning literally those which share our table. As omnivores, rats will eat almost anything. Thanks to mankind, rats have enjoyed thousands of years of relatively trouble-free living. Since the days when the Egyptians worshipped the cat-goddess Bast, praying for her to kill the rats which ate their food supplies in the royal granaries, rats have lived largely off the fruit and particularly the waste of man's labour. They prospered in the overcrowded and insanitary conditions of the middle ages when a large reservoir of flea-infested rats spread bubonic plague throughout the land.

By the middle of the last century rat-hunting or rat-fighting in the rat pits of London had become a popular pastime. Both gentlemen and poor men in Victorian England would pay their pennies to watch this horrific sport, in which dogs ploughed into piles of rats trapped in a three foot pit, attempting to beat bloody records by breaking the rodents' backs in their jaws. While rat-fighting was in vogue, rats were hunted in great numbers. Famous ratcatchers like Jack Black, Ratcatcher to Her Majesty Queen Victoria, sold hundreds of rats each week to the tavern owners who held the rat-killing matches. But as their popularity waned, fewer rats were required; by the turn of the century the size of Britain's rat population had increased dramatically. Great rat marches abounded in the countryside, as thousands of rodents moved from one colony to another. Witnesses described rats attacking and devouring those unfortunate enough to be caught in their path.

The advent of two world wars, slum clearance programmes, property development and the closure of the great warehouses in our once-thriving docklands, had a stabilizing effect on Britain's rat population. Advances in technology and the science of pest control transformed the humble ratcatcher into a 'rodent control officer'. The new methods of treatment and more effective rodenticides began to reduce the number of rats in the UK and by the middle of the 1970s, it looked like the war against the rat had finally been won.

Ridding Britain of its rat population is not a national or local priority. While rats remained below the surface out of sight and mind in Britain's crumbling sewers, their numbers were of little interest to anyone. No major investment programmes have been undertaken for sewer maintenance or renewal and a golden opportunity to bait rats to near extinction has been missed.

It is ironic that, while the size of the UK's rat population has been growing significantly over the last three years, scientists have never known more about rats or ways of exterminating them.

Researchers discovered a sixth sense in rats – kinesthesis – an acute awareness and memory of position. They also found that rats communicate at ultrasonic frequencies and they suffer from neophobia – a fear of new objects. These discoveries all helped in the search for faster and more efficient methods of destroying rats. Anti-coagulants were specially formulated to accommodate the new demands of pest control legislation and poisons were manufactured which killed rats in a more humane manner. Scientific research has also put to rest many rat myths and legends. Rats do not work in close co-operation when stealing eggs from chicken houses, for example. One rat does not lie on the floor holding the egg while the other drags him along by the tail. Nor is there such an animal as a King Rat, vested with leadership powers at the top of the rodent hierarchy who is followed and obeyed by an underling class.

Rats are host to myriad deadly diseases. As animals which frequent insanitary places like sewers, water courses and rubbish tips, rats spread many infections by carrying them on their fur and paws. Scientists have increasingly blamed rats for the rising incidents of salmonellosis; many serious cases were reported this year which were caused by rats wandering over food surfaces and leaving smears of infection. Some bacteria are transmitted to humans through food contaminated by droppings or urine, without the rat itself ever being infected. Rats carry rabies, typhus, Chaga's disease, rat-bite fever, and many other diseases which they transmit directly to humans, or to other animals which in turn pass them on to humans.

In Britain this year more than fifteen people died and over a hundred were made seriously ill as a result of contracting leptospirosis or Weil's disease. It is a deadly bacteria carried in rats' urine and has killed both adults and children whose work or leisure activities brought them into contact with rat-contaminated water. Leptospirosis is a preventable disease and can be treated with penicillin but despite the increase in the number of cases, public awareness of it remains low and its symptoms are often spotted too late. Rats are the sole carriers of Weil's disease.

Yet, despite all that we have learned through scientific research into the life and habits of the rat, despite our improved methods of control and our knowledge of the new threat posed by Weil's disease, even despite our bitter experience of history, we continue to ignore the problem of rats.

Last year the level of rat infestations in Britain shot up by more than a fifth, a national increase of around twenty per cent. This year's figures are set to rise further still, with many local authorities reporting increases in their local rat populations which range from ten per cent to 70 per cent for the first six months of 1989. Some people have blamed our recent mild winters for the rising number of rats. The winter of 1988 was indeed mild, the second warmest for more than 300 years, and there is no doubt that with fewer young rats killed by the cold, an increase in the breeding capacity and therefore the size of the rat population was inevitable. The warmer weather is not, however, the main cause of our current problem. The milder climate has acted as a catalyst on a series of other factors and has made an already dangerous situation significantly worse.

There is no national body specifically involved with rodent control, no Government department with a brief or budget for monitoring and dealing with the national rat problem. All rodent control matters are dealt with at a local level. With the current financial squeeze on local authority resources, councils have had less money to cope with the rising number of rat infestations. Many local authorities are unable to provide an adequate rodent control service at present, let alone face the new demands of increasing rat infestations. Less staff and funding coupled with the increasing costs of sewer and land rat-baiting has led to cutbacks in many associated services, especially refuse and sewer cleansing. Sewer maintenance and baiting programmes have been set aside by those local authorities which lack the financial resources to fund them.

The water authorities have failed to invest adequately in their sewer systems or in preventative baiting treatments. Drains and sewage are an expensive business, and regular maintenance of our decaying and rotting sewers, which in many cases date back to Victorian times, has been seen as literally throwing money down the drain.

The political and moral discussions over the Government's policy of water privatization – from the ethics of privatizing a service of significant importance to public health, to the quality of Britain's water – have pushed the rat problem to the side. The new private companies which are responsible to their share-holders look more likely to give profits and dividends a higher priority than the loss-making activities of repairing and re-building sewers.

Rats, as we will see, have always been a menace and a danger to mankind. *Rats – the New Plague* will trace the history of rats on their chequered path from the plains of southern Asia to northern Europe, through the Black Death and the Great Plague, the perils of rat attacks in Victorian sewers and the great rat marches of the countryside, and right up to the present disturbing situation.

The middle chapters will examine the biology of rats, their breeding and feeding habits and the diseases which they carry, looking at the latest methods of rodent control, and at the existence of 'super-rats', those resistant to rodenticides. The increasing official number of leptospirosis victims represents just the tip of the iceberg, probably a tenth of the actual amount, according to some experts. The book will investigate the new threat posed by Weil's disease, and look at what is being done to control it.

The work will then shift to the front line in the war against rats and look at the work of the brave army of rodent control officers – the Jack Blacks of days gone by. We spend a day in the life of a modern ratcatcher and ask those in the field what is responsible for the current increase in our rat population. The City of Manchester and the London Borough of Barnet come under the microscope, as we investigate the problems rats pose to public health and private safety.

Finally, some surprisingly simple solutions to this serious problem will be proposed. *Rats – the New Plague* is an attempt to bring the present rat crisis into the public eye, to set it in its historical and biological context and to suggest possible solutions to the problem. The technical battle against the rat has already been won. We have an armoury of scientific methods and poisons and an expert army of soldiers at our disposal. What we lack is the determination to fight.

The problem of rats is a national issue concerning public health; its importance should not be underestimated nor relegated to a mere question of finance. The larger the size of the rat population, the greater are our chances of catching rat-borne diseases. We ignore rats at our peril, if history is to be the judge.

Charles Golding
London
Autumn 1989

Part One

HISTORY

1

Rats of Old

'Over you and me hangs the same dread shadow,' wrote Karl Gustav Anker-Petersen in his book, *The Menace of the Death Rat*, back in 1933. 'The rat has gnawed its way into the very vitals of civilization. He is laughing at man's puny efforts to check his all conquering progress. You and I may be his next victim. In my opinion, only a miracle is responsible for thousands of Londoners being alive at the present moment.'

Strong stuff indeed, from the man who was regarded as the rat expert of his day. His writing brimmed with horror stories of death and disease caused by 'the evil rat'. Paragraph after paragraph warned against 'the most intelligent enemy man had ever known'. Its 68 pages were individually titled in capital letters, each with an emotive heading: 'Their Cunning and Cleverness', 'Laughs at Poison', 'Disease-Ridden Devils' and 'The Crafty Rat'.

However, *The Menace of the Death Rat* should not be taken too seriously. Anker-Petersen's book was an hysterical tirade against what he saw as 'the greatest threat to mankind since the very dawn of civilization, that foul, filthy enemy, the rat' and not a serious look at Britain's rat problems. He ended with this grave warning:

Whether we be rich or poor, the rat holds out his vicious threat to all of us. If we want to survive, we must strangle the monster before he strangles us. We must fight, fight with all our united strength to

check and conquer this hideous and all powerful enemy. We must
begin the battle soon – at once – before it is too late.

The author was, of course, slightly biased – he had something
to sell. Karl Gustav Anker-Petersen was, for nearly thirty years,
manager of the joint Danish/British company known today as
Rentokil. In those days it was but a small concern called the
British Ratin Company which had been set up by a Danish
operation, the Bakteriologisk Laboratorium Ratin. The book
was a sensationalist attempt to get British companies to take the
rat problem seriously and to employ, presumably, the Ratin
company to do it. It contained a wealth of exaggerations and
inaccuracies – and every old wives' tale he could unearth. But
Anker-Petersen was right about something. The rat *has* gnawed
its way throughout history, and has been with us since the dawn
of time.

Tracing the precise origin of the rat is not easy. Most
authorities agree that *Rattus rattus*, the black rat, was to be found
in the Ancient World. Its first home was probably southern
Asia. Rats are said to have travelled to Egypt, then the
Mediterranean, finally reaching the countries of northern
Europe. There is evidence of rats in ancient Indian scriptures,
of 3000 BC. Ishwar Prakesh, the editor of *Rodent Pest Management*, writing in 1988, claimed that rodents have been dreaded
for thousands of years in India, and that there are countless
references to granaries being ravaged in ancient Indian
scriptures:

> Oh Ashwani, kill the burrowing rodents which devastate our food
> grains, cut off their heads, break their necks, plug their mouths so
> that they can never destroy our food. Rid mankind of them.

Rats certainly are well represented in early Indian culture.
Every Indian god has an animal or creature associated with it.
Vishnu, for example, travels with a bird. There is also a god
with an elephant's head, called Ganesha, who rides on the back
of a rat.

In Rajasthan today, in the village of Deshnoke there lies a

temple built to the local goddess Karni. The temple is teeming with rats, which are allowed to run freely around the building and its grounds. The temple's architecture was specially constructed to incorporate rat holes and rat runs within the structure. Legend has it that the villagers were suffering greatly from the plague, and called on their goddess Karni for help. She advised them to build a temple to her, in which the rats could run around freely and be happy, and so not harm the people. It seemed to do the trick. By removing the rats from the village, the plague disappeared, possibly making it the first link in history between the plague and rats, hundreds of years before its discovery in the rest of the world. Tourists today are even encouraged to feed the rats, and nets are suspended over the rat runs to protect them from harm.

Cargo ships were universally held to be responsible for unwittingly carrying rodents around the world in their supplies of wheat, spices and grain. There were two possible periods in which black rats could have arrived in Egypt.

During the 3rd millenium BC, there was much seaborne trade in the Indian Ocean and the Persian Gulf which would have provided ideal transportation for rats out of India. The ships which carried cotton supplies from the Indus valley might have been one source; the large granaries of Mihenjo-daro and Harappa would have attracted the rats in large numbers. They could also have come during the Ptolemaic, Roman and Coptic periods. Under the Ptolemies, for example, Egypt was well known as a great trading nation, acting as a convenient stop between India and the Mediterranean. Spice, teak and sandal-wood were transported to Egypt, along with rats in great numbers.

The ancient Egyptians worshipped many gods, including a cat-goddess called Bast. It is said that cats were revered by the Egyptians because they kept down the number of rats in the city and especially the grain stores. The royal granaries were essential to the Egyptian economy and to the life of the people; they acted as emergency supply stores when drought and famine hit the land.

When the Romans occupied Egypt between 30 BC and AD 641 the country's trade with south Asia continued, and the rats arrived in ever growing numbers. The spice trade was at its peak, and most spice came from India. The stomach of a mummified cat, recently discovered in a Roman building dating around the first to the second century AD on the Red Sea coast at Qusir al-Quadim was found to contain the remains of six black rats. The town was a mercantile community, with many links with India.

Once rats were established in Egypt, they spread with the help of Roman sailors and merchants. Following its victory over Egypt, Rome imported grain from Alexandria directly to its ports in Italy. Rats came too. There is archaeological proof of the existence of rats at this time from the site at Pompeii, dating back to the second century BC. The eastern Mediterranean and southern Gaul lay wide open to ships which were often as large as 1,000 tons. From there it is likely that rats moved to northern Europe especially Germany and Switzerland. Again there is archaeological proof of this. They travelled to Spain and North Africa too, following the trade routes. Rats were beginning to populate the world.

Up to 10 years ago, received wisdom had it that the black rat arrived in Britain around about the time of the Crusades, between the 11th and 12th centuries AD. Historians generally held the view that rats returned with the navies of the conquering crusaders, alongside the booty plundered between Jerusalem and Britain. Detailed dates and places were given, with ultimate 'proof' resting on the absence of any archaeological evidence to the contrary. All that changed, however, with the discovery of rat remains at the excavation of a Roman well in Skeldergate, York.

In 1979, James Rackham, a Senior Research Assistant in the Department of Archaeology at Durham University, published an article identifying the skeletal remains of *Rattus rattus* at the site at York, dating it to the 4th century AD. A rodent skull was recovered from layer 2404, and identified as that of a black rat on the basis of the cusps of the teeth and the parietal ridges. The skull was found to be largely intact. Sieving through deposits below the skull level, two further finds were discovered, indicating that there had been at least two rats in the Roman well. The specimens were found in the lower part of the well, which was lined with timber and constructed between the second and third centuries, but which was thought to be still in use a century later.

Many similar claims that skeletal evidence of rats in archaeological sites is sufficient proof that they existed at that time, have been discounted on the grounds that rats probably entered the sites at a much later date, burrowing down and dying in the older levels. In the York case, however, the possibility of the entry of rats, either as live animals or as remains in the backfill of the well, was excluded. The excavators found no signs of burrowing, or holes allowing rats to penetrate through the five metres of deposit. The well had been effectively rat-proofed. In any case, as we shall see, black rats, unlike their brown counterparts, are unlikely to burrow to any great depth.

Rackham noted that there had been little direct evidence of the occurrence of rats in Britain prior to the 14th century, apart from the York findings and similar evidence from a dig in Southampton. He cited the writings of Giraldus Cambrensis at the end of the 12th century as the earliest convincing reference to the existence of rats. Cambrensis wrote about a man being 'persecuted by rats' and refers to rats being 'expelled' by St Yvorus from a district called Fernigenan. Rackham's paper on the discoveries at York struck at the heart of the widely accepted notion that the earliest rats came to Britain as stowaways on the returning ships of the Crusaders in about 1095, 1147 and 1191 AD. The York evidence, as we have seen, dated rats back to Roman Britain, around at least the 4th century AD.

Further evidence came to light as recently as 1983, proving the presence of *Rattus rattus* in Roman Britain beyond any doubt. A paper presented by Armitage, West and Steedman on the findings of a dig at Fenchurch Street in the City of London revealed the bones of black rats which dated to the middle of the 3rd century AD. Their paper proved that a well established population of black rats lived in the City of London around the 3rd century AD, linking their findings with the recovery of other bones from a 4th century site at Crosswall. The paper admitted that it was still not possible to ascertain exactly when and how black rats arrived in Britain, but suggested that they may have been brought in as a consequence of trade with Europe – either from the Rhineland or the Mediterranean.

So, the myth of our 11th century rat invasion – swarms of black rats arriving in Britain for the first time with the returning Crusaders – can finally be put to rest. Not all stories, however, surrounding the humble rat are as easy to disprove.

Probably the most famous legend in history concerning rats is still an open case today. Some say it happened, others say it did not, still more argue that something indeed took place, but not what the story books tell us. Just what did happen to the rats, the children, and the Pied Piper of Hamelin some 700 years ago is still anyone's guess.

Most people have a rough idea of the tale. A little village somewhere in Germany had a rat problem. They called in a strange ratcatcher who, using his magic pipe, sorted out the town's problem by playing the rats a merry tune which made them follow him to a watery grave in the local river. The city councillors refused to pay the piper's fee so the disenchanted ratcatcher piped himself, and the city's children, out of the village forever. So ends the story. But the debate over the facts continues.

It is probably Robert Browning's version of the Pied Piper of Hamelin which most people in Britain know today. Written in 1849, it included all the elements of popular appeal. It was a story about innocent victims – the city's children, greedy councillors, a mysterious, magical but vengeful piper and, above all, horrible, vicious and troublesome rats.

> Rats! They fought the dogs and killed the cats,
> And bit the babies in the cradles,
> And ate the cheese out of the vats.
> And licked the soup from the cook's own ladles,
> Split open the kegs of salted sprats,

Made nests inside men's Sunday Hats,
And even spoiled the women's chats,
By drowning their speaking,
With shrieking and squeaking,
In fifty different sharps and flats.

Browning wrote a wonderfully entertaining story in rhyme but took advantage of his poetic licence. Hamelin Town (which everyone but the British spells Hameln) is in Lower Saxony, not Brunswick, as Browning had it. There is no proof that the town was suffering from a plague of rats nor a shred of evidence to link any swarm of rodents with the arrival of a pipe-toting stranger or with the documented disappearance of the town's children. Even the date that Browning gave us – 1376 – is open to question. In short, Browning was no historian.

Christopher Nicholson, in a witty 'Whodunit?' examination of the case of the Pied Piper, listed a range of possible explanations. The man in the brightly coloured set of clothes might have been the leader of a tribe of gypsies who abducted the town's children. Perhaps they were killed in a fearful accident or were poisoned, or died in the local marshes. Did they leave on a Children's Crusade for Jerusalem or were they struck by the bubonic plague? The only real clues, according to Nicholson, are several ancient documents and a few wall inscriptions.

A chronicler by the name of Richard Verstegan in a book with the curious title of *A Restitution of Decayed Intelligence in Antiquities*, wrote a rough history of the events, in 1602. He gave the date of the Pied Piper incident as the 22nd July, 1376. Verstegan's account is probably the one on which Browning based his poem.

But in the *Book of Ordinances and Statutes* in Hamelin, which goes back over six hundred years, the entry which noted the disappearance of the town's children is dated at 1351. Another ancient manuscript, the *Luneburger Handschrift* of 1450 dated the events at 1284 and put the number of children missing at 130. That ties in with the original inscription on a house in Hamelin built in 1602. The house is situated in a road called

Bungelosenstrasse – the Street of No Drums. (Legend has it that after the children vanished into a hole in a mountain or hill a law was passed prohibiting the playing of drums, pipes or any musical instruments.) A gold inscription on the outside wall states that on the 26th June, 1284, a piper in multi-coloured clothes led 130 children past the village's place of execution and out of the city where they were lost forever.

Browning's poem referred to a church with a stained glass window which commemorated the event. To make it all the more confusing, there was said to be a church in existence in 1300 with a window depicting the legend which has since been destroyed.

Where the children went is another mystery. The French and German Children's Crusades took place around 100 years earlier, but the youngsters may nevertheless indeed have followed a charismatic leader with a pipe and gone to the Holy Land with a later crusade. They might have migrated towards eastern Europe in common with nearly a million Germans at that time or died in an outbreak of plague which affected the younger and weaker members of the community. The German historian Werner Ueffing claimed in 1983 that Hamelin was hit by the bubonic plague in 1351. The victims' bodies were buried in a hill and it was possible that a man with a pipe walked in front of the heap of corpses ready for burial to warn people away. It looks like we will never know.

In all of these explanations nothing definitely links the missing children to rats. Plagues of rodents were common at the time, and there could have been a demand in Hamelin for a good ratcatcher to purge the town of them. He might well have used a pipe – British ratcatchers have done so in the past. It seems most likely that the event of the missing children happened first as there is no mention of rats until the 16th century. From then on, both stories were told as one legend from one generation to the next, and the myth was born.

The final word on Hamelin must go to the resourceful management team at Rentokil which, believe it or not, in 1965 became the first official ratcatchers to the town of Hamelin since the Pied Piper. The company signed a contract for rodent control with the City Fathers who passed a special bye-law fining anyone who failed to notify the council of a rat infestation the sum of DM 500. Rentokil contracted to rid Hamelin Town of all its rodents for ten years, but insisted on being paid quarterly, in advance. That news prompted the *Evening Standard* to write, with apologies to Browning,

In Hamelin Town, in Brunswick, by famous Hanover City,
The Mayor announced to the Corporation:
'We have a moderate infestation
   Of rats.'
Since nobody paid the Piper's bill,
We'll have to call in Rentokil.
But the firm knew the tale of the Piper's pay.
And they went to the Mayor with this to say,
'Give us the fee that we determine,
Before we rid you of your vermin.'

Rentokil was paid regularly and the parents of Hamelin slept soundly at night.

2

Bubonic Plague

The deadly association between rats and the bubonic plague has always aroused a morbid curiosity in mankind. Rats have carried with them a reputation for spreading diseases; but nothing was more terrible than the infamous Black Death which ravaged Europe from the middle of the 14th century, leaving millions dead.

It is likely the plague has been with us since early history, existing in areas around the central Asiatic plateau, where it probably originated. Plague pandemics usually follow a pattern of three stages. They begin with a short and violent burst of outbreaks of the disease with a high rate of mortality. There then follows a long endemic period of occasionally erupting epidemics, which gradually die away. Finally, after what might be several hundred years, the plague vanishes.

There have been three plague pandemics. The first was thought to have started in Arabia, and to have moved to Egypt in 542 AD. This is the outbreak which ravaged the Roman Empire of Justinian. The plague travelled throughout Europe to England, where it was known as the Plague of Cadwalader's Time, and went on to Ireland around 664 AD. The second pandemic was known as the Black Death, which was introduced into Europe through the ports of the Crimea around 1347 and culminated in the Great Plague of London, in 1665.

It is said that we are still suffering the third pandemic – which started in Yunnan, then crossed to Bombay in 1896, where it claimed the lives of nearly six million Indians. Plague was reported in Suffolk, Britain, in 1910, but fortunately it failed to take hold in the rest of the country. This pandemic has yet to run its entire course.

In the *Textbook of Medicine*, Professor Fred McCrumb defined the plague as an acute or chronic disease of wild rodents transmissible among themselves and to humans through the bite of infected ectoparasites. Most experts agree that only *Rattus rattus* – the black rat – has the biology and behaviour suited to spreading and supporting bubonic plague amongst humans, although there have been isolated outbreaks arising with other species of rodents. The brown rat – *Rattus norvegicus* – makes its entrance into the history books at a much later date and, although it is capable of carrying bubonic plague, it was not to blame for the Black Death.

The mechanics of transmission of bubonic plague from black rat to human being are as horrifying as the symptoms themselves. The deadly plague bacillus, *Yersinia pestis*, is carried in the rat's blood, where it multiplies. The flea, *Xenopsylla cheopsis*, takes in the bacteria when it feeds on the rat's blood. In many cases, the bacteria divides itself up and quickly forms a solid mass in the flea's stomach. The flea is then known as 'blocked', as no more blood can enter its stomach. The flea suffers acute hunger, and looks for a warm blooded animal on which to feed.

When it finds a new host, often a human, the flea attempts to feed again, but finding no room left for its meal, stretches its gullet to the maximum in an attempt to take in more blood. The elasticity of the gullet causes it to contract, and some of the blood is forced back into the wound. It is at that point that the bacillus gets into the bloodstream. Another unpleasant detail is that, when a flea feeds, it automatically defaecates, depositing bacteria into its minute faeces. The victim, thus irritated by the bite, often scratches it unintentionally, so rubbing the organisms into the wound.

The plague symptoms were accurately observed by the chroniclers of medieval times and seem to have been roughly the same as those of today's sufferers. The incubation period can be anything from two to eight days. The victims suffer from 'a swollen and dropsical mass of inflamed lymphatic glands' often the size of an orange and display dusky blotches under the skin. The plague bacilli concentrate in the lymph glands, causing them to expand into large and painful swellings in the groin, armpit or the neck. The name bubonic comes from the Greek word *bubo*, meaning groin. Graham Twigg, a zoologist and biological historian, noted that marked skin manifestations – dark patches of internal haemorrhages – were common during the Black Death but are a rare thing this century. The line in the children's rhyme 'Ring a ring of roses . . .' is said to refer to the dark body patches of haemorrhages around the victim's waist.

Bubonic plague was slow to set in, but once it had taken hold, the victim might be dead within a week. According to modern research, if the patient can endure around five days of intense pain and fever, and the bubo breaks down and suppurates within a week, the patient will survive. Twigg notes that in some instances, where the bacteria gets into the bloodstream directly, avoiding the defences of the lymph nodes altogether, septicaemia sets in quickly and the patient dies after a short illness. In cases of primary septicaemic plague, death is swift indeed, with the patient retiring to bed feeling quite well, and dying within a few hours during sleep. These cases are, however, infrequent.

During an attack of bubonic plague, organisms find their way down to the lungs, where they can produce a third form of the disease – pneumonic plague. The children's rhyme previously referred to contains the line 'Atishoo! Atishoo! We all fall down!' which is thought to refer to this type of plague. Pneumonic plague centres specifically on the lungs, which produce sputum infected with the plague bacterium. Unlike the bubonic form which is transmitted to man through infected rat fleas, pneumonic plague is highly infectious and is spread by a simple cough or sneeze transporting hundreds of tiny infected droplets into the air. If untreated, pneumonic plague results in a death rate of 100%.

What happened all those years ago to cause hordes of rats to suddenly leave their homes in Asia and settle in all parts of the world remains a mystery. Suggestions have ranged from starvation – local food supplies were perhaps inadequate to support the entire rat population – to natural disasters affecting their colonies, like flood or possible drought. All that is certain is that at the middle of the 14th century, Britain and the rest of Europe provided ideal conditions for both the black rat and the spread of plague.

There has been some controversy over whether the rat was entirely responsible for the outbreak of plague at that time. It has been proven that the flea, *X. cheopsis*, can live for up to a month without either a human or rat host. This could have enabled it to live in a ship's cargo travelling hundreds of miles to Europe, before passing the plague directly to man, by biting the unfortunate sailors who unloaded the merchandise. The theory has been impossible to prove, and the general view held now is that rats were indeed mainly, if not exclusively, responsible for bringing the Black Death to Europe.

The Black Death arrived in Sicily in 1347, in boats returning from the Crimea. The plague followed the trade routes, and it is likely that infected fleas and rats travelled in the cargo holds of merchant ships, particularly those carrying wheat and cereals. The historian Philip Ziegler, in his excellent book on the Black Death, noted that it was quite common at that time for rats to be host to more than one flea, the average being around three fleas to a rat.

The period 1340 to 1400 saw the beginnings of an important transformation of Britain from a medieval to a modern society. Trade was on the increase, alongside agriculture and manufacturing, largely as a result of the breakdown in the feudal system. Increased economic freedom led to more personal enterprise as serfdom gradually disappeared. Greater freedom and more money led to a significant increase in the population.

There have been several attempts to calculate the exact population of England at that time; the general consensus is around 4.2 million, although the figure ranges several million on either side. Just under 12% of the population lived in towns and cities. London was the biggest and most prosperous, boasting around 16,000 households in 85 parishes, approximately 70,000 people in all. The next largest city was Norwich, with 13,000, followed by York with 10,000 inhabitants.

When the Black Death struck, it killed nearly a third of the country's inhabitants in just over two and a half years, a total of 1.2 million people. The growth in the size of the population had led to widespread overcrowding and insanitary conditions and whilst these two factors alone did not cause the plague, they facilitated the rapid spread of the disease.

Fourteenth century Britain abounded with black rats. The country could not have deliberately created a more ideal environment in which the rats could thrive. When the bubonic plague reached England in 1348, it found no shortage of victims and although the plague did not recognize class barriers, the labouring poor, living in squalid and unhygienic conditions with inadequate sanitation, were easy targets. Tainted water, foul air, rotting refuse and food remains dumped in streets with open sewers – if they existed at all – all took their toll on the general health of the population. Dysentery and diarrhoea weakened the victims' ability to withstand an attack of plague.

Black rats are not great burrowers. They also like to live near food and water supplies. Thatched roofs and hedgerows provided secure homes, and the abundance of food, rotting in the open or poorly stored in sheds and outhouses, provided the rodents with a stable environment and so their numbers increased. It was estimated that every household in England was home to at least one family of rats.

Rows of hovels were crammed together often blocking out the light. These poorly ventilated dwellings were home to dozens of people; it was not uncommon in the country for families to share their accommodation with farm animals like chickens and pigs. The majority of hovels were made of timber and clay, easy materials for rats to gnaw into and build nests. As Ziegler wrote, 'The medieval house might have been built to specifications approved by a rodent council as eminently suitable for the rat's enjoyment of a healthy and carefree life.'

The plague swept through the Dorset ports down the south coast of England through Devon and Cornwall, its main thrust leading up the Thames Valley towards London. By March 1349 it was spreading all over England, from Norfolk and Suffolk, Cambridgeshire, Hampshire and Surrey, Warwickshire and Worcestershire and towards the northern counties. By 1350 all of Britain was affected. Although the rich left the larger cities in an attempt to avoid the plague, no place remained untouched. People who were thought to be carrying the plague – strangers or visitors new to an area – were turned away and prevented from entering a village or town. Rats however escaped detection by following ditches and streams and swimming over moats and rivers, bringing with them their deadly fleas.

London in 1348 suffered from an acute overcrowding problem. With an estimated population of around 70,000, it was not uncommon for 12 people to be found sleeping in one room. Open sewers and cesspools, slops and garbage in the streets made the city an insanitary place. Ziegler noted that the King had complained by letter to the Mayor in 1349 about the unbearable filth in the City and asked him to do something about the human faeces in the streets and lanes which 'poisoned the air'. A pocket full of posies might have helped the judges disguise the smell, but the royal request was to get rid of it altogether. The Mayor was too busy to respond; with more than 20,000 corpses to bury, the collection and disposal of bodies had become a priority in plague-stricken London.

The initial outbreak of the Black Death in 1348 was without doubt the worst that the country was to experience. In just over two years, a third of the inhabitants of Chaucer's England were wiped out. Chaucer himself, who was born in 1340, escaped the plague which he referred to as 'the great pestilence' in *The Pardoner's Tale*. In the three hundred years following Chaucer's time, Britain suffered from numerous violent outbreaks of plague in different areas – most commonly at ports and towns near rivers – but the Black Death never again swept the whole nation as it did in 1348.

The plague remained endemic in London under the Lancastrian and Tudor kings. According to historians, celebrations in the city over the accession of James I were cut short by an outbreak of plague which killed more than 30,000 people. During the Civil War of 1642 to 1646, plague was particularly bad in the South and West of England, although the city of Chester lost nearly a quarter of its inhabitants. But despite these deadly outbreaks, the population of London continued to grow, and by Tudor times births outnumbered deaths.

The Plague of London in 1665 marked the last serious outbreak of the Black Death. It was not the worst attack, but following nearly 30 years of plague-free living, it must have seemed so to the city's inhabitants. In the first week of December 1665, two Frenchmen died in a rented house in Drury Lane, the first recorded victims of bubonic plague in London for three decades. Six weeks later, in February 1666, another plague death was recorded and by May, the epidemic was in full swing. Samuel Pepys wrote in his diary on the 7th June that he saw three houses that day near to where he lived bearing red crosses on their doors warning people that plague victims dwelled within. King Charles, seeing the writing on the wall, moved his entire court out of town. Two thirds of the city's population then followed suit and made for the countryside, heading up the river Thames and consequently taking the plague with them.

The country braced itself for what it thought would be another violent epidemic on the same scale as 1348, but it never came. Ironically enough, London was to be saved by a freak accident, a disaster which had been waiting to happen for nearly a century, and which threatened to destroy the entire city. The Great Fire of London in 1666 reduced much of London's timber and clay buildings to charred rubble; but up in flames along with the slum dwellings and hovels went the homes of millions of rats and their infected fleas.

The process of rebuilding the city changed the black rat's

domestic environment, as timber and clay gave way to bricks and mortar. Rats found it harder to make nests and move freely from one feeding site to another because the new buildings made harbourage difficult and restricted movement. For the first time in its history in England, the black rat found its way of life under attack. But if the pressure was on at home, an even greater challenge to its existence was waiting just around the corner. In under sixty years, the black rat would do battle for its very life, not against man, but against another species of rat. And it would lose.

The brown rat is thought to have originated in China and Siberia, although the exact location is not known. Some historians have suggested that it came from Chinese Mongolia, and have pinpointed the region to the east of Lake Baikal or along the Caspian Sea as its home. There is little more information on their entry into Europe. According to Pallas writing in 1831, masses of rats were seen swimming across the Volga in 1727 following an earthquake, and swarming into Astrakhan, moving westward across Russia. They arrived in Britain in the late 1720s, and were referred to as the Norway rat, because they were supposed to have travelled from the East on Norwegian timber ships. Historians and biologists alike have given precise dates for the brown rat's appearance in the countries of Europe: Italy and France 1750; Sweden 1790; Switzerland 1809 and so on. Of course, as Mark Hovell pointed out, writing on the subject in 1924, the reporting of vast hordes of rats crossing rivers by witnesses – however reliable they may be – does not prove the exact date of the brown rat's arrival in a particular country. 'Prussia 1750; Norway 1762; Faroe Islands 1768 . . . as if they had dates stamped on their passports.'

As we shall see when we examine the biology of a rat in more detail, *Rattus rattus* never stood a chance against its more adaptable cousin, *Rattus norvegicus*. Brown rats are considerably bigger than black rats, weighing up to a pound and measuring an average of nine and a quarter inches. Brown rats are ferocious fighters too and with size on their side, had no problem in driving the black rats out of their former strongholds. The biologist Graham Twigg noted an interesting report made about the time of the brown rat invasion – around 1768 – by the official ratcatcher to a European princess. The ratcatcher gave a good example of how the 'Norway' rats had driven out the black rats and killed them. He had trapped some black rats at the top of a house, and placed them in a large cage overnight, along with some brown rats which he had caught in the cellar. By the morning, he wrote, the brown rats had eaten all of the black rats.

Brown rats are much more able to survive cold and harsh weather conditions, they can produce large litters more quickly and are not as particular over what they eat; brown rats literally eat anything. The black rat with its long tail is a born climber, preferring to live in hedgerows and on the top floors of buildings. The brown rat, however, is a born burrower, fully able to climb but choosing to stay at ground level, or below. It therefore survived the transformation from thatch to lead and tile roofs, from timber to brick; in short, it was the more adaptable of the species.

Within a few years, the brown rat had become established in England as an unrivalled nuisance. It was such a hated pest that it became known as the Hanoverian rat, a reference to the unpopular King, George I. The Hanoverian monarch, who had come to England in 1714, was so loathed by Roman Catholics, Tories and Jacobites alike that it was claimed he had brought the brown rat with him on his royal ship. Just over a century later, black rats had virtually disappeared from the UK, with only a few surviving colonies scattered around the country. They had been driven out and exterminated by their own kind.

Very little has been recorded on rats in Britain between the years 1760 and 1850. Placing the odd report of rodent infestations here and there into one giant jigsaw gives us some picture. The newly arrived species of brown rats appear to have multiplied and prospered, invading all areas of human life. In 1783, Thomas Swane, the man in charge of catching rats on board the ships of His Majesty's Royal Navy, killed 2,475 rats aboard the *Duke* and 1,075 on the *Prince of Wales* using a paste made of white arsenic, meal and sugar.

The biologist Zinsser noted that at the siege of Paris in 1871, when people trapped behind the barricades were suffering from starvation, rats were eaten without apparently causing any harm. Similarly, in the French Garrison at Malta in 1798, rat carcasses fetched a high price as food. The ancient sailors of Spain and Portugal made no secret of eating rats on long voyages and it was recorded that the starving sailors aboard Magellan's ship, which made an ill-fated attempt to circum-navigate the world, bought rats from each other at one ducat apiece. It should be noted, however, that the one animal which is excluded as a potential meal in the SAS Survival Manual, even under the most desperate circumstances, is the rat.

By the middle of the nineteenth century, brown rats were well established in Britain. For most people they were a source of disease or death, a constant pest and nuisance and a sign of poor hygiene, filth and poverty. But to a minority they were a source of livelihood and entertainment. The Victorians reduced everything into trade and profit and rats were no exception.

3

Amusing Rats

If images of Victorian sewers flowing with offal from slaughter-houses, pavement dirt of every kind, entrails from butchers' shops, stable dung, pig sty refuse and dozens of dead cats and dogs is likely to disturb, then the subject of rats in the Victorian era should be avoided at all costs. This is not a chapter for the squeamish. The history of rats in 19th century Britain takes the reader into those places which people would rather avoid, unless something goes wrong. Sewers and rats were not popular topics of conversation in genteel Victorian society, and the sewer workers, flushers, ratcatchers and sellers of rat poison went about their business largely unnoticed.

Yet there was another ghastly aspect of the small industry which had grown up around rats, which was very public indeed – but not, once more, for those of a sensitive disposition. For just one shilling, the Victorian gentleman could enjoy – if enjoy is the right word – a macabre night out at the sport of rat killing, where he could watch people betting on dozens of ferocious dogs as they ripped their way through hundreds of live rats in a large pit. A typical example of such an evening's entertainment was recorded by Henry Mayhew, the man who wandered amongst London's poor in the 1840s observing their work, living conditions, and general way of life. Mayhew's colourful first-hand account published in 1851 of a night's rat killing could rival any scene created by Charles Dickens.

Men of every grade of society gathered at a particular public house which was well known for holding a weekly event. Most spent their time smoking, drinking, and talking about dogs. Dogs were at the very heart of a rat fight; it was the skill and stamina of each hound which would determine how many rats he could kill within a set time. Bets were taken in the same way as horse racing today, and the winning dog, the one who destroyed the largest number of rats in the shortest time, received a special collar and a purse for its owner. Many spectators brought their own animals. Bull dogs, Skye terriers and English terriers were the most popular breeds.

On the tavern wall hung a print of 'Wonder' Tiny, a dog of five and a half pounds, who achieved fame when he killed 200 rats in one hour. This was not a record however. Billy, the tavern's favourite, it was claimed sent 500 rats to meet their maker in just five and a half minutes. According to the proprietor's son, rat fighting was a bloody business, and many dogs suffered appalling injuries as a result of savage rat bites to the mouth. Blistering ulcers often resulted from such wounds, and they had to be lanced so that the unfortunate creatures could be sent back into the pit several times each week.

At nine o'clock, on the evening of Mayhew's visit, the tavern's shutters were closed, and the gas lamps above the pit were lit. The pit resembled a small circus ring about six feet in diameter, surrounded by an elbow high wooden fence. The rats, several hundred of them, were kept in a wire cage at one end. Mayhew wrote that '. . . the rusty cage looked alive with a crawling black mass.' Before the main event, it was the custom for dogs to be bought and sold. Their comparative strengths were tested by throwing them into the pit with a dozen larger rats and watching how they performed. Mayhew witnessed several demonstrations and sales, and when the trading was over, the dead rats were swept up into a pile in the corner, ready for the Grand Match.

Fifty rats were to be killed in the first bout. They were hand selected from a large chicken basket and dropped into the pit, where they scurried into one corner on top of each other so that they resembled a live heap of hair sweepings and reaching a third of the way up the side of the wall. A smell rose from the black mass like that from a hot drain, no surprise really, since most of the rodents had been caught in London's sewers and water ditches. A bull terrier, nearly mad with rage at the sight of the rats, was let loose in the pit and rushed headfirst at the mound, burying its face in the pile and pulling out just one rat, which it tossed into the air, breaking its neck.

Very soon, several dozen more lay dead or bleeding on the white floor of the pit, which had now become stained with blood. One rat refused to let go of the dog's nose, so the terrier charged towards the wall, dashing against it and leaving 'a patch of blood as if a strawberry had been smashed'. At the end of the contest, ha'pennies were thrown into the ring for the dog's second, the pit was swept down and more rats were selected for the next round of sport. The evening's entertainment came to a close at midnight, and the Victorian gentlemen went home greatly amused.

With over 40 pits in London and bill boards regularly advertising special events which promised the slaughter of more than 500 rodents at one time, the demand for live rats was enormous. In the age of the ratcatcher perhaps the most famous was the Rat and Mole Destroyer to Her Majesty Queen Victoria, Mr Jack Black.

Jack Black had been ratting for 35 years. He had been taken on by the Royal Parks supervisor at a time when the open fields were heavily infested with rats which undermined bridges and gnawed at drains. His fee of six pounds per annum was soon changed to threepence per rat, a penny less than the going rate in the country. Black soon heard that the Royal ratcatcher, Mr Newton, was dying, so he hurried off a letter to the Royal Ordinance applying for the job. Within a few weeks, Black had been appointed the Queen's ratcatcher, whose responsibilities included all the army barracks in London.

The job was not full time, however, and Jack Black supplemented his income by travelling around London as a freelance ratcatcher, advertising his skills at street markets. He wore a showman's costume of white leather breeches, a green coat and scarlet waistcoat and a gold band around his hat. His crowning glory was his leather belt, which he wore proudly across his left shoulder. Black's original belt was painted with four white rats, but his best effort was a true tribute to his profession – four metal rats sitting proudly on a black leather sash, in the same pose as one might expect from a pride of lions. Black made the decorations himself, using a plaster cast taken from a dead rat.

The belt was designed to draw in the crowds and it worked. Jack Black would stand inside a huge cage crammed with rats and allow them to crawl about his feet and up his legs. He was often seen playing with some of his specially trained pet rats, which he would keep inside his shirt or allow to run up and down his arms and balance on his head. The purpose of his performance was to sell his rat poison, which Black often tested out live on his rats, to prove how quickly it worked. At the Tottenham street market, Black sold packets of his 'composition' for one shilling each. On some days he took more than four pounds in sales. Great acts of daring were often needed to spur people on to buy his concoction. On one occasion, Black pulled seven live rats out of the cage with one hand at the same time and one of Jack's friends even put the head of a live rat into his mouth.

Ratcatching was undoubtedly a dangerous trade – Jack Black came near to death three times in his 35 years in the business – but considering the number of risks he took, this figure is amazingly low. On one occasion a rat bit clean through a bone in his finger, snapping it in two. Bone breakers were the worst sort of bite, according to the Royal ratcatcher; the pain was unbearable and the wound would bleed for hours afterwards. Many bites resulted in sores which festered. As a precaution against this, Black cut each bite 'clean out with a lancet'! But despite the pain and suffering, the Royal ratcatcher enjoyed his work, was proud of the job he did – and recalled how people were always pleased to see him. On one occasion, Jack Black was called to a house in Hampstead where two children had woken up in nightclothes 'covered in blood'. An examination revealed rat bites to their hands and feet. Jack Black put down his deadly composition, and killed all of '. . . the spiteful fellows with serpent-like heads . . . the blood rats!' The family were pleased, according to the ratcatcher, paid him well and affectionately called him 'Mr Ratty' when he saw them again.

Jack Black, ratcatcher to the royals, showman, maker of poisons, keeper of dogs and ferrets, in fact, a real-life Jack-of-all-Trades, also bred fancy rats, which he decorated with ribbons and sold mainly to ladies. As we shall see later on, this tradition is kept alive in Britain today by the National Fancy Rat Society.

Black clearly enjoyed an ambivalent relationship with the furry creatures. Although most of his life was spent trapping and killing rats, he must have had a side which actually liked them; he let several nest in his shirt during the day, and spent hours training them to do tricks. But before we get carried away with the notion that here was a man who was really devoted to rats, remember this: Jack Black told Henry Mayhew that he had happily eaten several rats, which he found to be 'as moist as rabbits and as nice!'

Mayhew interviewed a few other ratcatchers in Victorian London, mostly from Brill Place, Somer'stown. With the possible exception of those of Jack Black, ratcatchers' stories, it appears, should be treated in the same way as those of fishermen. Exaggeration would certainly seem to be the order of the day in the case of one interviewee, who told Mayhew that he had taken a bet which involved competing with a dog in the number of rats they could both kill with their teeth. The ratcatcher claimed to have beaten the hound and won three sovereigns to boot.

Ratcatchers made their money in three ways: selling poison; freelance rat killing to rid buildings of infestations and catching rats for resale to the tavern keepers for rat fighting. Street sellers of rat poison usually carried pictures of evil looking rodents and were accompanied by the ratcatcher's friends, the terrier and the ferret. Most of the home-made poisons used to treat rat infestations, for both domestic and commercial use, were arsenic based, and lethal to cats and dogs too. Mayhew wrote: 'When a ratcatcher is thus accompanied, there is generally a strong aromatic odour about him, far from agreeable, owing to his clothes being rubbed with oil of thyme and aniseed mixed together.' It was claimed that this smell was so attractive to rats that they would simply rush out of their holes towards it, another fisherman's tale as far as rats go, most likely. The smell of aniseed has however proven highly attractive to dogs, and may have helped terriers locate their masters more easily.

The rats which were needed for the rat fights were trapped live and kept in large cages. Once a rat's nest had been located, all of its possible exits would be blocked except one, which would be open on to a wire cage. Ferrets would then be sent in to chase the rats out.

Catching rats in the sewers was an altogether more dangerous business. Victorian sewermen worked in truly awful surroundings; the absence of legal restrictions governing the dumping of filth and toxic waste meant that some sewers were little better than covered cesspools. Refuse from gas works, breweries, chemical and mineral factories mixed with ashes, old kettles and pans, rotten vegetables and other food waste, jam jars, pitchers, old mortar and bricks, timber and rags as well as human effluent was freely dumped out of sight below ground. The sewer hunters, as the workers were popularly known, made light of their working conditions, but most admitted that their chief fear was that of a rat attack. In the better constructed sewers, rats were quite rare, but the older and more decayed sewers were home and breeding ground to packs of aggressive rats.

Although Mayhew found both black and brown rats for sale in the streets most sewer workers had never seen a black rat. Black rats were largely unknown in the underground sewers. Some of the men described vicious attacks by rats which they claimed were the size of kittens, and which flew at them while they worked. One flusherman whose job it was to keep the sewers clean of obstructions by flushing away blockages, remembered that, just before Christmas in 1847, he had seen cartloads of drowned rats floating along the river Thames, by the West Strand shore and the gates at Northumberland Street. The flusherman then swept them into another underground stream.

The sewers under the meat markets were the worst for rats, a flusherman told Mayhew. This was due to the large amounts of animal offal swept into the drains at Newgate, Whitechapel, Clare and, of course, Smithfield. These sewers attracted rats from all over the system, drawn by the smell of rotting meat, and served as a perfect breeding ground, offering an undisturbed existence with a plentiful supply of food, fresh water and warmth. In 1860, a slaughterhouse for horses in Montfaucon, France, which processed around 35 carcasses each day, was the subject of a test to find out how many rats lived in the country by observing the numbers feeding at that site. It was said that the bones of the dead horses were being picked clean overnight by rats. Horse meat bait was set in traps at various parts of the building, and cages were established where the rats could be trapped, killed and counted. Over the period of a month, more than 16,000 were caught, 2,650 on the first night alone. It is unlikely that this was a representative sample of the country as a whole, but it gave some indication of the enormous amount of rats in the area.

Back in Britain, the rat continued to prosper, and by the turn of the 20th century, rats were a serious problem in the town and country alike. In the Port of London, a new method of ratcatching successfully netted 48,000 rats in 1901. Thirteen years later, the average catch had been reduced to 12,500 and the science of rodent extermination had begun to succeed in the cities at least. The country proved a much friendlier place for brown rats; it was darker, less hostile and had more feeding opportunities. Changes, which are a common characteristic of town living, do not occur with the same frequency in the country, which therefore offers rats a more reliable food supply. Brown rats lived mainly in farmland in storehouses, outbuildings and other structures with little or no mortar foundations. Cow sheds, poultry houses and pig sties were particularly popular, as they offered a new colony food, shelter and warmth.

During the early part of this century, talk of large rat migrations was fairly common. Rats share the habit of moving en masse with several other species of rodent, such as the lemming. It should be remembered that rats arrived in Europe as a result of an original mass migration from the Asian plateau. In 1900, sporadic reports of smaller rat migrations appeared throughout the country. Rat marches were particularly common in Wiltshire, Berkshire, Hampshire and most of the corn growing counties. Several circumstances prompted these events, the most common of which was that the rats' food supply had become too small to sustain the growing colony.

Many old wives' tales have grown up around rat marches. Some suggest that some mystical feeling of impending doom prompted a would-be rat council to organise a mass evacuation of the colony, as in the beginning of Richard Adams' novel, *Watership Down*. Examining a few historical accounts of early twentieth century rat marches reveals a common story line – 'vast armies of rats marching in disciplined columns through the open fields in pursuit of a particular barn or field led by a distinct leadership'. In truth, it is unlikely that rats marched with such precision and determination under a line of command. Observers probably witnessed a mass of rats crossing a field or road, driven on by nothing more than hunger or thirst, sent packing by a rat hunt, a field fire, or the destruction of their nesting grounds and following no specific order or direction, in search of a new place to live. Other descriptions were more likely due to overactive imaginations than scientific observation – king rats, rat generals or any other hierarchical levels of rat society are popular myths, as we shall see in the next chapter.

One horrific story of a rat march which ended in the death of a tramp came to light in 1978, when it was published in *Essex Countryside* magazine. The June issue reported a first hand account of a 77-year-old woman who witnessed the event as a teenager during the First World War. The rest of the article is a poor rag-bag of assorted myths and misinformation about rats, but the story of Miss Patricia Vernon, a land girl on a farm in Nottingham in 1914, has a ring of truth about it.

Vernon told the reporter that she was picking potatoes in an open field at the time, when she saw what looked like a dark, ground-level cloud coming towards her. The farmer she was with warned her to lay down, cover her face, and stay perfectly still. Vernon followed the man's advice and what she described as a fifty yard long column of rats squeaked their way past her, without causing her any harm. She said she was frozen with fear, and remembered it well.

Not long after, Vernon stated, a similar incident occurred, but this time an old tramp stood between the girl and the rats. Vernon said she shouted the farmer's instructions to the old man, but he failed to hear her and was eaten alive by the horde, screaming horribly, while Vernon stood by, helpless to act. When it was safe, she ran over to the poor fellow, finding nothing but the remains of his clothes, boots and bones: 'the rats had eaten every part of his body,' she said. Was Vernon's story the result of an imperfect memory adding fiction to fact or did it really happen the way she told it? Rats have been known to go on the march and they will also eat human flesh or any other meat if given the chance, but it is uncharacteristic for rats to attack humans, unless they are cornered and deprived of an escape route. Experts have yet to agree on the truth of Patricia Vernon's tale.

By the start of the First World War, rats were becoming a major national problem, as evidenced by the large number of ratcatching advertisements in the popular press and in the range of specialized books on sale. There were dozens of remedies on offer, with enough expert opinion to sink a battleship. Of the many compounds and rat traps which were available, few survive today – the backbreaker being the notable exception – but some should be praised for their ingenuity at least, if not for their efficiency.

One such idea was developed by B. Winstone & Sons, in Shoe Lane, London. 'Ratsticker', according to its label, was a safe, non-poisonous ratcatching compound selling at one shilling a time, or a half-a-crown for a large tin. The concept was simple; the compound, as its name suggests, was a kind of super glue which was boiled in a pan of water and poured onto a 15 by 10 inch piece of cardboard. Cheese was then placed in the middle of the sticky trap, which was left next to the rat run. Dr Howarth, the City of London's medical officer, testified it was successful in catching as many as 80 rats at a time. They died quickly too, according to the doctor, from shock! A Brer Rabbit-styled trap with a death-by-heart-attack ending was doomed to failure from the start and went the same way as 'Sidebotham's Steel Rat Trap', 'Klearwell's Mixture' and 'Magic Paste'.

Do-it-yourself ratcatching books were so popular that they won their own library category – 'Rat Destruction'. Amongst the titles included *The Prevention & Destruction of Rats* by Sergeant-Major E. B. Dewberry, RAMC, *The Rat Problem* by W. R. Boelter and *The Rat and How to Kill Him* by Alfred E. Moore, referred to as *The Official Handbook of the International Vermin Repression Society*, a group which seemed to have disappeared as quickly as it was founded. That none of these tomes made it to the scientific best seller list is no great surprise since most were spoiled by their subjective personification of 'the evil rat'. They did however contain some useful hints on contemporary ratcatching for the absolute beginner.

One book in particular deserves a mention: *Rats & How to Destroy Them* by Mark Hovell. Although it was written over 60 years ago, it is still used as a reference book by some specialists today, and in one London Borough at least, it occupies pride of place in the rodent section of the pest controller's personal bookshelf.

By the end of the First World War pressure mounted on Parliament to pass legislation forcing individuals to treat their own rat infestations. Up to that point, local authorities lacked the power to act against landowners who neglected their properties and allowed them to become breeding grounds for rats. Rat killing for sport was a thing of the past and the era of state rodent control was about to begin.

4

Of Fancy Rats and Kings

No century has had a greater effect on the rat's domestic environment than the current one. Despite the gradual changes in the town and country over the last five hundred years which have forced rats to adapt to new feeding and living conditions, the twentieth century has had the most radical effect on the rat's relatively peaceful existence. The intervention of the state in matters of rodent infestation through local government bodies and supporting legislation, coupled with the growth of new technology – inventing more specific poisons and new methods of extermination – have made it the era of rodent control. As we shall see in more detail later in the book when we look at the progress of control legislation, the end of the First World War saw the state take its first serious steps towards solving Britain's increasing rat problem.

On 28 August 1918 the Rat Order came into effect. The measure was passed by the Food Controller under the Defence of the Realm Regulations after Government reports revealed that rats were costing the taxpayer an estimated £52 million per year in damage to crops, buildings and livestock. The new law gave powers to local authorities to go into land infested with rats and exterminate them, where the occupier or the owner had disregarded previous warnings on rodent control. Legislation also empowered the authorities to recover their operational expenses. The measures marked a turning point in the history of the British rat. It was the first time the state had intervened in the rodent problem, which it had previously regarded as a purely domestic issue. The Rats & Mice (Destruction) Act 1919 codified most of the details of the Rat Order, paving the way for the Prevention of Damage by Pests Act of 1949 which is still in force today. But if the government passed national laws controlling rats the responsibility for their implementation passed downwards. It was local government which had been given the job of policing the level of rat infestations, a legal obligation of crucial importance today.

Plague outbreaks, a constant reminder throughout Europe of the dangers of an uncontrolled rat population, continued to flare up around Britain. In 1924 the first serious outbreak in the British Isles this century struck in Glasgow, Scotland. There were 36 plague cases directly linked to rats and 16 people died.

During the thirties and forties, especially during the Second World War, rats continued to be a major problem. Very little has been written on them during this period although it was noted that, as an island country at war suffering from acute food shortages, the significant destruction of Britain's very limited supplies of food by rats was a major worry to several Government Ministries. Rats living in corn ricks were found to be a particular problem. The problem was so great that a group of eminent biologists met in Oxford to study the effects of rats on the food supply, looking for better methods of control.

After 1945, new poisons were discovered and methods of baiting improved and local government played a more active part in rodent control. These methods proved such a success that it looked as if the war against the rat might be finally over. Of course, there were still incidents of appalling infestations. In 1949 for example the last of the grain windjammers, a great ship called the *Pamir*, was treated for rat infestation. It had been moored at Penarth while being used as a giant floating food store. Despite the ship's rat guards and hawsers, an astonishing 7,894 rats were removed. In general, though, slum clearance programmes and new building methods made the rat's life more difficult in cities and towns.

A popular myth in circulation at the time concerned the

existence of Rat Kings. This legend seems to be as old as the history of the rat itself and many people still claim to have seen Rat Kings at work to this day. It is suggested that in some rat colonies, older, perhaps wiser rats have some sort of authority over the younger members of the group. Several people say they have seen particularly large rats leading convoys of others away from a warehouse or a condemned property just before it was to be destroyed. Eyewitness accounts of so-called rat marches have claimed the existence of a specific leader. But biologists have put these stories into the same category as other 'confirmed' accounts of supposed rat co-operation. They are simply not true. Despite popular belief, rats do not help one another; the story of two rats seen working together to steal an egg from a hen house, with one hanging onto the egg and the other dragging him along by the tail, has no scientific foundation whatsoever.

The original expression 'Rat King' was first used by medieval Germans – *Rattenkonig*. There is a similar term in French, *Roi des Rats*. Both terms describe people who live on the backs of their fellows. Martin Hart, in his book on rats, cites Konrad Gesner who uses the term 'Rat King' in his *Historia Animalium* (1551–8), 'Some would have it that the rat waxes mighty in its old age and is fed by its young – this is what is called the Rat King.' But years of research have failed to identify one colony of rats with an identifiable leader or demonstrate any cases of individual or systematic help for OAP rats from their younger colonists.

Another 'Rat King' which has surfaced from time to time is the collection of several rats tied together by their tails. Hart says there are several theories behind this, the most unlikely being that they were so formed in the womb. Another improbable suggestion is that the stronger rats may have tied the older and weaker rats together to make a nest. Hart's most favoured theory is that the rats could have become accidentally entangled when frightened as rats do have the habit of wrapping their tails around branches when under stress; the knot could have been tied by sheer chance when the huddled rats were suddenly surprised.

The earliest known depiction of a Rat King was published in 1564 in Antwerp by Johannes Sambucus in his *Emblamata*. The most interesting story of a recently discovered knot of rodents is cited by Hart, a fascinating account of a Dutchman who found seven attached rats in 1964. The Dutch farmer killed a black rat in his barn, only to discover when he tried to pull it from a pile of bean sticks that it was joined to six others. He killed all of them and on closer examination found the Rat King to consist of two males and five females of roughly the same age. The knot included most of the tail of one rat, and just the tips of the rest – the tails appeared swollen and dented where they had been tied together. An X-ray revealed several fractures in the tails themselves, and in some of the vertebrae. All these findings suggest that the tails had been knotted together for a considerable time.

Despite the facts, the mythology of Rat Kings – rodent rulers or groups of rats knotted together – is alive and well, fuelled by the occasional miraculously unsubstantiated eyewitness report. Less than a decade ago it was claimed that a ratcatcher by the name of H. G. Wood of Battersea killed the legendary 'monster' rat of London, the Giant Rat of Tooley Street. An article in a local magazine described Mr Wood as a man who believed that such great rats bred exclusively in their own family clans which stuck together from generation to generation under the leadership of the oldest, wisest and biggest rat, and that they lived more or less charmed lives! While large and aggressive rats do exist on this planet, 'royal' rats with magical powers live strictly on the moon, probably dining on green cheese.

Not all ratcatchers, however, tell tales to rival fishermen. Brian Plummer, a man who caught his first rat when he was eight and has hunted them for sport ever since, takes rats very seriously. In his book, *Tales of a Rat-Hunting Man* published in 1978, Plummer recounts some of the most bloody stories ever written about rat-hunting; his personal experience as a modern day ratcatcher is unrivalled, and, oddly enough, although his writing is peppered with blood and gore, it is a remarkably entertaining read. Plummer knows his subject well. There are several practical chapters in this macabre autobiography on the use of ferrets and terriers in ratting, and some history too. The occupational hazards of rat-hunting do not seem to have prevented Mr Plummer from enjoying his sport; like Jack Black, he suffered several nasty wounds from both rats and ferrets, but they failed to put him off his chosen hobby.

One of his most hair-raising experiences involved a rat hunt in Sutton Coldfield, near Birmingham. Plummer had planned to 'bag' the rats live as he hunted in a hen house. It was stiflingly hot, and he was wearing a baggy knitted sweater. An old buck jumped out of his grasp, up his arm and into his jumper, dashing around and making catching him a very difficult process. Plummer eventually got the rat by the tail, and dragged him out, biting and scratching all the way. It was a horrible experience, even for the hardy ratcatcher, and Plummer admitted that the event deterred him from hunting for the rest of the night.

Brian Plummer's successful days as a ratcatcher were unique; by the sixties and seventies, most untrained individuals with little or no technical expertise in rodent control found they could no longer make a living from ratcatching.

Eccentric individuals, however brave and locally experienced, could not compete with large companies like Rentokil, offering the systematic extermination of all pests, including rodents, and armed with the latest advances in safer and more effective toxic treatments. Jack Blacks finally gave way to rodent control officers. It is recommended in Britain today that rodent sightings and surface infestations should be reported to the local authorities or to a private firm of pest controllers where trained experts will deal with the problem. The same is not true, however, in China, where the effects of centuries of plague outbreaks have promoted the rat to the public status of a national menace.

Despite the great number of plague deaths throughout China's history, and the potential risk that rats still pose there, the country has a bizarre do-it-yourself policy towards rodent control. Chinese citizens are encouraged to kill the rats themselves when they encounter them. A campaign of wall posters and leaflets instruct the proletariat, including children, to kill any rats as soon as they are found and not to wait for any expert help. Individual ratcatchers still survive in China, an occupation dating back thousands of years to the first recorded mention of rats in the country. Written in 600 BC, one of the poems in *The Book of Songs* pleads: 'Big rat, don't eat our grain!' At Wuxi, a village near the mouth of the Yangtse River, a local ratcatcher sits today outside the doorway of his house, playing his pipe. The pipe is meant to attract rats – a Chinese version of the Pied Piper – and as proof of his success, a semicircle of dead rats is laid out at his feet.

But the Chinaman is not alone in using this method as a way of attracting potential customers. Thousands of miles away in France, for example, one Parisian shopkeeper displays, alongside poisons and treatments of every description, several stuffed rats hanging from traps in his window. It is a gruesome and unexpected sight in one of the most romantic and sophisticated cities in the world.

The 20th century has seen fundamental changes in methods of rat extermination in the UK. Legislation, the development of poisons and the national and systematic treatment of rodent infestations have all had a radical effect on the rat's living conditions. Jack Black might be dead and his trade demoted to a subdivision of a pest control operation, but his spirit is still with us, at least according to Malcolm Cleroux.

Mr Cleroux is the President of probably the strangest club in Britain – the National Fancy Rat Society. The society, which takes itself very seriously, traces its roots to Jack Black, who trained some of his rats for exhibition and sale to the gentry. Fancy rats were particularly popular with some Victorian women, who kept them in parrot cages as pets.

In 1901 the first national Mouse and Rat Club was established in London and by the First World War, pet rats were common to all classes; there are several stories of soldiers keeping them in their uniforms in the trenches of France. By the end of the war, however, their popularity had begun to wane and most of the organizations died out. Throughout the 1970s, renewed enthusiasm for pet rats led to the demand for a national club where breeders and buyers could proudly display their pets, and in 1976 the NFRS was formed. The object of the society is to study and show off fancy rats. Mick Brinklow, an area co-ordinator with the society, describes fancy rats as very friendly, clean animals which come in all sorts of varieties like the Siamese and the Berkshire. Brinklow, who keeps 40 rats for breeding purposes at his home in Kent, claims he was the first man to breed an all-white rat with black eyes, as opposed to the usual red. According to the society's enthusiastic president, it boasts more than 400 paid-up members, including 'someone high up at the BBC, James Allcock, the TV vet, and even the daughter of one of the directors of Britain's largest rat extermination company!' The society's own magazine, *Pro Rata*, is taken by some universities; its articles discuss general health issues and genetics.

The National Fancy Rat Society strongly defends itself against charges that it is a collection of eccentrics playing with potentially dangerous animals. The NFRS stresses that its members do not breed wild rats, and its literature issues warnings that membership will be withdrawn from anyone attempting to introduce wild rats into the group. Malcolm Cleroux recommends rats. He says they make good pets, especially for children, and respond well to training and learning tricks. He claims to have seen hundreds of rats pass through the society's hands without a single incident of illness. But there are many who feel that the potential risk of catching an infection from a rat which might have been bred from wild stock far outweighs the possible enjoyment of keeping a rat as a pet. Experts have warned of several dangerous diseases borne by rats, including rats which originated from the laboratory; there is no such thing as a pure breed.

Brinklow told a local newspaper earlier this year that he hoped people would come along to one of his rat exhibitions and realize what attractive animals rats really were. 'Rats have got a very bad public image', he complains, 'mention them and most people go "ugh"!'

If people do go 'ugh', it is not without good reason; rats have brought plague and death to millions through the centuries. They are dangerous creatures who have done nothing good for mankind; they continue to steal his food and destroy his property. But rats in Britain pose a more immediate danger. As we shall see in the next section when we examine the biology of a rat, more than half of today's rats are infected with a potentially deadly disease, which, if their numbers continue to rise unchecked, could pose a serious threat to public health in the UK. If we are to learn from past experience, rats – fancy or wild – should still be taken seriously.

Part Two

BIOLOGY

5

The Rat Race

While examining the biology of a rat, its habits and diseases, its patterns of reproduction and feeding, it is easy to see why it is such a detested creature. Rats are extremely effective at carrying diseases and infecting humans and it is important for mankind to fully understand their biology if they are to be efficiently controlled.

Rats are defined as a pest worldwide, and are responsible for the direct destruction of millions of pounds of crops and farming produce each year. Rats spoil twice as much grain as they eat by contaminating it with their droppings, thereby making it unfit for human consumption. They gnaw through power cables, gas pipelines, lead piping and even bricks, and cause floods, fires and explosions the economic costs of which are impossible to estimate. More dangerous by far are the many lethal diseases which rats have spread to humans – typhus and bubonic plague, and more recently, salmonellosis and leptospirosis.

Yet, were nature handing out awards for survival in the mammal category which numbers more than 5,000 different species, then rodents such as rats would undoubtedly be the prize winners. There are over 2,000 types of rodent in the world, the largest being the *Capybara*, which weighs in at nearly one hundredweight, and the smallest the *Pygmy mouse*, which manages just one third of an ounce.

Rodents get their name from the latin verb *rodere*, meaning to gnaw. Gnawing is an essential part of a rat's life. Unlike most other mammals, rodents have no canine teeth, and rely on a single pair of incisors. These chisel-shaped front teeth grow continuously throughout a rat's lifetime. Should a rat damage its teeth, or fail to find enough hard material on which to gnaw, the incisors grow in a complete circle and have been known to pierce the jaw.

Rats are also extremely efficient feeders, eating virtually anything, animal, vegetable, fresh or foul. Each year, rats eat and damage an estimated 20% of the world's crops. In Asia alone they consume 50 million tons of crops per year, enough to feed over a quarter of a billion people.

As we have mentioned previously, the two main species of rats are *Rattus rattus* and *Rattus norvegicus*.

Rattus rattus is better known as the black rat – a confusing name since black rats can often be grey-brown or tawny in colour. They are also known as ship or roof rats because they were said to have arrived on the shores of the British Isles by ship, and preferred living in the roofs of houses and other high places to the ground. Black rats have pointed muzzles, large ears, protruding eyes and thin tails, which are longer than their heads and bodies.

Rattus norvegicus came to Britain in the middle of the eighteenth century – people presumed it came from Norway, hence its popular name, the Norway rat. It is also called the brown rat, the sewer rat or the common rat. Common it certainly is, since it has all but extinguished black rats in Britain. Brown rats have blunt muzzles, small and furry ears and are grey-brown in colour. Their tails are shorter than their heads and bodies in cold climates and longer in the tropics. Unlike black rats, brown rats can live in town just as easily as in the country, and prefer living at or below ground level.

Rats are very aware of their environment and enjoy the security of familiar objects, therefore preferring to stay put once a suitable home has been found. As we have seen, great rat marches are rare events, which only occur when prompted by extreme changes in home circumstances, such as floods, famine or the destruction of an entire colony. Rats generally move to find food and water, shelter and breeding partners, usually staying within the home range of the colony. When disturbed, however, they can run far distances, and experiments have shown that rats which have been trapped and released elsewhere will move up to four miles before settling down in a new habitat.

Brown rats tend to move downwards; they burrow into the earth, and are attracted therefore to the sewers. This pull in a downward direction is known as negative geotaxis. Black rats, on the other hand, look for places in hedgerows and roofs, preferring to live well above the ground. Black rats are better climbers than brown as they are slighter and more agile, but it is as well to remember that all rats can and do climb. As A. P. Meehan states in his excellent book on rats and mice, 'When conducting a survey for a rodent control operation, it is imperative to look upwards and sideways, as well as downwards.'

Apart from burrowing and climbing, rats swim well too. Brown rats have stayed afloat for 72 hours in tests before beginning to tire and, as we shall see in the next few chapters, rats are more than capable of swimming up waste pipes and entering houses through the 'S' bends of toilet systems.

Brown rats are also first rate jumpers, proportionately better than Olympic athletes. Brown rats can jump upwards more than 77 cms, and horizontally from a standing position more than 120 cms. Black rats do slightly better than this; one rat managed to leap 150 cms diagonally, from a shelf to a light fitting.

Rats have five senses, much the same as humans, but with differing degrees of importance. Touch is essential to a rat, which tends to be a nocturnal creature spending most of its life in dark places. The rat's major organs of touch are its *vibrissae* – whiskers – which are very sensitive. Laboratory experiments have shown that rats prefer close touch with objects. Once familiar with a particular route or run, rats will use it again and again and well-established runs are generally easy to spot in rough mud, as they are marked by a smooth surface, the result of many scampering feet. This is one of the first signs a rodent controller looks for when investigating reported infestations.

A fascinating sense of touch possessed by rats is known as *kinesthesis* – muscle sense – which is the ability of an animal to learn all the parts of its environment by bodily contact alone. According to Meehan, it is based on 'the subconscious memory of sequences of muscle movements' which practically means that rats will 'learn' a particular route by its very twists and turns, without relying on other senses, such as sight and sound, which might indicate a change in conditions. Rats tested in this area, which originally ran around obstacles in order to avoid them, continued to do so, even after the objects were removed. The opposite was also found to be the case. This information is particularly important when laying poison and traps for rats.

Smell – *olfaction* – is critical to rats. Most odours influencing reproduction are present in their urine. Rats display extreme behaviour when reacting to certain smells. They 'freeze' when they detect cat odour, yet they show no signs of reaction when a cat comes into sight but are denied its smell. It is based on this principle that some companies now market smells which are supposed to 'frighten the rat to death'. Generally, rats can distinguish between individuals of the same species through their smell. It is interesting to note that while the smell of a human has little effect on wild brown rats, the odour from cow, horse and pig urine produces a very positive reaction – rats are very attracted to it.

Sight, as mentioned before, plays a small role in the day to day life of a rat and although they are probably colour blind, rats have an ability to recognize shapes and movement in the dark.

Rats have, on the other hand, a strong sense of hearing and are able to pick up sounds which go into the ultrasonic frequency, out of range of the human ear. Rats can also produce high frequency emissions for communication between each other, with each sound lasting less than 300 milliseconds. Newborn deaf pups produce ultrasound to attract their mothers' attention and it is sometimes used for echo location, in a similar way to bats.

Finally rats respond to four tastes – just like humans; bitter, salty, sweet and sour. Rats will eat anything that man does, and just about anything besides, as we will see when looking at their feeding habits.

Rats are born survivors, and propagate in large enough numbers to guarantee their future. An important reason for controlling rats is their ability to breed quickly and in large numbers. Ironically, it is exactly this characteristic which makes them so popular with scientists looking for subjects on which to experiment in laboratories. This fecundity is typical of most small animals which are used as prey by several predatory species. Most have relatively small lifespans, short gestation periods, rapid sexual maturation and polyoestrus breeding cycles.

A healthy pair of rats can produce over 100 direct offspring in one year, and be responsible for a further 800 descendants over the same period. Of course, this assumes an average litter of ten, ideal conditions for survival and that all their offspring will reproduce. But even if the true figure were nearer a quarter of that number, it stresses the importance of acting immediately to control new rat colonies before they become established and their numbers rapidly escalate.

Rats reproduce in a manner similar to many mammals. The male has testes, a penis and internal accessary glands which produce spermatozoa. When sexually mature, the male's testes hang outside his body in scrotal sacs. The female's reproductive organs are ovaries, located in her abdominal cavity. During heat, ovulation takes place in the female, when the egg is released in the vicinity of the fallopian tube, from which it makes its way to the uterus or womb. When the male deposits his sperm, fertilization takes place in the uterus, and the eggs are implanted there.

The oestrous cycle, which is common to most mammals, takes just four to six days in female rats, compared with 28 days in the case of humans. Brown rats are sexually mature in eight to twelve weeks after birth – black rats take an average of four weeks longer – and in normal conditions, the female gestates or carries her young for around twenty-one days. Although the young rat pups are born blind and hairless, within a week they are able to see, although they are still reliant on their mother's milk. Rats continue to grow even after reaching sexual maturity. The female rat, unlike many other mammals, restarts her oestrus cycle soon after giving birth to her litter. In other words, rats can give birth every twenty-six days. At any one time, around 20 % of female brown rats are pregnant, with litter sizes varying between 9 and 11 pups.

As with most animals, the most important factors affecting reproduction are temperature and nutrition. Both of these elements, as we shall see later in the book, are at their optimum point in Britain today, where we have experienced some of the mildest winters on record and a huge increase in food refuse on the streets.

There is much evidence to suggest that temperature is the major factor influencing the reproductive capacity of rats, once adequate food and shelter have been secured. Indeed, amongst outdoor rats, the reproductive capacity of individuals is increased during the warmer months. Brown rats do not like to reproduce in very cold conditions. Competitive behaviour is an important factor too, and rates of reproduction are affected by crowded conditions. When one species declines another steps in to fill the void.

The lifespan of a rat is surprisingly short. It is estimated that wild rats live for less than eighteen months, but views on this vary. Some claim the average life expectancy of a town-dwelling rat to be just one year. On some farms, only 5 % of the original brown rat population survived one year, and for black rats the number was only 3 %.

Today, of course, it is human activity which has become a major factor in determining the size and lifespan of the rat. Both in the town and country, it has been man's technological advances which have radically affected the natural growth in the size of the rat population as, through his use of rodenticides and traps, man has managed a significant level of control over numbers.

The question of lifespan is important in relation to controlling the size of a rat colony. If births were to suddenly exceed deaths, then the population would rise dramatically. Such is the current problem in Britain, where our cold winters could usually be relied on to kill off large proportions of the younger rat population. Our recent milder climes are producing the twin effects of allowing breeding to continue throughout the winter months, and ensuring that the average number of surviving pups in litters rises.

The most common man-made method of control throughout the world today is poison. But even this is becoming less effective. In the developed world where so much food waste finds its way onto the streets, the competition is intense between this appetizing refuse and special rat-killing dishes containing poisons. It is no easy job to attract rats to a particular meal, when discarded bags of chips and half-eaten hamburgers are such tempting alternatives.

Rats are omnivorous, eating anything on offer, but like most mammals, they require a balanced diet. Rats must drink water daily to survive, and unless their food source is very moist, they prefer to live with a fresh source of water nearby. They have been known to lick the condensation from window panes of barns on farms during the winter months in order to survive.

It is a widely held view that although rats will eat just about anything they come across – they are quite happy to perch on a pile of rotting material in a sewer and pick up and eat the scraps floating past – rats prefer cereals if given the choice. Although there is no such thing as a universally accepted bait, experts observe that cereals make the best base in most cases.

Young rats eat more than their parents do, who eat about 10 % of their body weight each day. They have no typical style of eating; some rats choose to sit back on their haunches, holding the grain between their paws and rotating it as they nibble. Others put their noses directly into the food to eat, while some rats use their paws to scoop up the food instead.

As we shall see in the following chapter, a well-fed and heathy rat population inevitably leads to an increase in its size, with alarming implications for public health.

6

Weil's Disease

One of the most frightening aspects of the growth in the size of Britain's rat population is the potentially huge increase in the cases of Weil's disease which may result. Between 50 % and 70 % of all rats in the UK carry the disease which is caused by a bacterium known as *leptospira ictero-haemorrhagiae*. Weil's disease may be passed on to man directly through contact with rat's urine or indirectly through contact with infected water.

Over the last five years, incidents of leptospirosis have been slowly rising. Figures for 1988 are beginning to show the effects of the jump in size of the rat population, with nine deaths from Weil's disease, an increase of 33 % on the previous year.

Weil's disease was named after a German doctor who, in 1886, discovered the combined symptoms of jaundice and renal failure. Dr Weil named it leptospirosis – *lepto* in Greek means minute worm, and *spira* means coil – an exact description of the organism. The leptospira is very mobile and capable of travelling at a fantastic rate, allowing it to easily enter the human body through any mucous membrane, such as the eyes and mouth, or through the most tiny of cuts or abrasions.

Leptospirosis is one of the major bacteria with which rats can live quite happily, unlike salmonella which causes the animal to be ill. Rats contract leptospirosis while still pups, and carry it in their kidneys. They excrete the bacterium through their urine into the environment, where it can live for up to 45 days in fresh water or in damp conditions, such as muddy river banks. The leptospirosis bacterium cannot, thankfully, multiply outside of the rat.

Groups most at risk have historically included farm workers and those whose jobs bring them into regular contact with rats. But with the increase in popularity of water sports over the last ten years, Britain's canoeists, who number around one million, have found themselves at risk, representing nearly half of those affected by Weil's disease in 1988.

The onset of symptoms may be gradual or sudden and display all the signs of a severe influenza. The victim suffers from an acute headache, combined with muscular pains and a general feeling of weakness. The symptoms also include a high fever, shivering, nausea and vomiting. Conjunctivitis develops, commonly seen as 'piggy-red' eyes and it is this which can first indicate to a general practitioner that the patient may be suffering from more than a heavy bout of 'flu. At this stage the disease can be treated successfully. Danger increases when these symptoms go unrecognized.

The first phase of Weil's disease lasts for about a week during which time septicaemia occurs but there are no signs of leptospira organisms in the urine. If the GP is not aware that the patient may have come into contact with rat's urine or rat-infected water through work or leisure activities, she is unlikely to correctly diagnose Weil's disease. Jaundice occurs in around 40 % of cases; should it fail to set in within two weeks, however, it is only a mild attack and the symptoms rapidly abate. The outcome can still be favourable if left untreated – even in the more severe cases – with the fever falling and a general improvement in the patient's condition, although the jaundice may remain for another two weeks.

In acute cases left untreated, jaundice becomes intense and passing urine difficult, with solid bodies forming in the kidney. Skin blotches, nose bleeding and blisters may develop and there is a rise in the pulse rate accompanied by a fall in blood pressure. The patient falls into a critical condition. Renal failure and kidney damage occur simultaneously and death might follow. Around 15 % of Weil's disease victims died last year.

In the majority of cases in this third acute stage, however, recovery takes place which may be accompanied by the development of antibodies in the bloodstream. This period varies, lasting months in some cases, with the patient experiencing a second rise in temperature during the third week of illness, before full recovery.

Although no one disputes that there is a significant increase in the incidents of leptospirosis in the UK, just how sharp the rise has been depends on which set of figures are used (see Table 1). Government figures have been criticized by those working with the disease for being 'artificially depressed'.

The official numbers are compiled by the Communicable Disease Surveillance Centre in London from the legal notifications sent by doctors to the Registrar General and the Chief Medical Officer. This practice began in 1968 when leptospirosis became a legally notifiable disease.

These statistics are too low. There are two possible reasons for this. Firstly, busy general practitioners and hospital doctors do not always report the disease when they discover it. This is true of most notifiable diseases. In the case of measles, for example, very few doctors report every case they come across. It is not just a state of apathy, we are assured by the good doctors, but that the fee until recently paid by the government to doctors for the report was lower than the cost of the postage stamp required to send it! Secondly, by the time leptospirosis has been diagnosed, the patient has usually been given a first 'misdiagnosis' which will remain as the official diagnosis on record.

The laboratory figures originate from the dozens of medical laboratories throughout England and Wales. These statistics are again artificially low, for similar reasons as the official figures. Some laboratories simply do not bother to send in a tally of any of the diseases they isolate and identify. Most labs work from week to week, and a diagnosis may not be made immediately, so it is carried over to the report for the following week or weeks. Many get lost in this process.

The figures confirmed by the Leptospira Reference Unit are

Table 1: Annual cases of Leptospirosis in Britain 1978–1988

	1978	1979	1980	1981	1982	1983	1984	1985	1986	1987	1988
Official Notifications (I)	11	18	13	23	14	23	19	31	22	25	36
Laboratory Reports (II)	22	43	14	30	31	81	53	72	45	57	99
Leptospira Reference Unit (III)	65	55	48	72	62	120	90	107	72	68	109

Sources:
(I) Notification to Communicable Disease Surveillance Centre.
(II) Laboratory Reports to CDSC.
(III) Annual cases confirmed by PHSL Leptospira Reference Unit.

without doubt the most accurate compilation; they simply express a more complete picture of the laboratory statistics.

Dr Sheena Waitkins, the director of the Leptospira Reference Unit in Hereford, the country's national leptospirosis research centre, believes the other figures are unreliable.

'My professional attitude is that we're wasting time and money on making diseases legally notifiable at all,' she says, pointing out that countries like America, France and Germany all use their own laboratory figures, compiled by research institutions like the Unit. Dr Waitkins goes further, however, saying that her own figures reflect only ten per cent of what is really going on, just the tip of the iceberg, 'I can only diagnose what I can see', she adds, 'there's a lot more besides.'

Dr Waitkins is perhaps the most respected authority on leptospirosis in Europe, if not the world. Following her first degree at Glasgow University, she took a PhD in the treatment and testing of gonorrhea, and became known for co-inventing the Flynn & Waitkins gonorrhea test. Her researches in the subject of medical testing took her to Denmark, but she was invited back to the UK by the government-funded Public Health Laboratory Service to work as a diagnostic scientist on streptococci.

The PHLS has a specialized unit in leptospirosis covering all of Great Britain, and operates with international commitments to the World Health Organisation, and to members of the Commonwealth countries.

Dr Waitkins has strong views on Weil's disease.

It is not the coming of a second plague; the majority of people don't get leptospirosis. However, the problem is that those who do are not being recognized in time. They become seriously ill, and die. And because the disease is sometimes incorrectly diagnosed, we are only seeing a tenth of what is really going on.

Her research work in Hereford is thorough and impressive. It involves diagnostic and testing methods – trying to understand the disease itself. Until the late 1970s, the test used was known as the Complement Fixation Test (CFT). This relied on a simple blood test to see if the sample contained antibodies, showing that the organism had entered the body. There are two types of antibody: IgM, produced in the first three to seven days of illness, and IgG, produced between ten and fourteen days. The CFT was only capable of detecting the IgG antibody.

The early 1980s saw a new test introduced which could find the IgM antibody. The test, called Enzyme Linked Immuno Assay (ELISA), could detect antibodies in the first three to seven days of the illness. The ELISA test became available and Dr Waitkins quickly adapted it for use with leptospirosis, where time was the key factor. 'The important thing here was that, particularly with the most severe varieties of the disease, early detection meant the possibility of early treatment and cure.'

Britain is now the only country in the world which still operates CFT testing in some of its public health laboratories. The ELISA test is quite expensive, costing between 50 % to 60 % more than the CFT. But analysts say the old test is labour intensive, which in the long term is more expensive. With the PHLS performing 35,000 tests each year, the economic and practical advantages of ELISA testing are obvious.

According to Dr Waitkins, between 10 % and 15 % of patients die unnecessarily of leptospirosis each year in Great Britain, and yet the treatment is very simple: high doses of penicillin, dialysis and good medical support. She is convinced the best way to fight the disease is by educating the public to be aware of situations which might expose them to danger, both at work and during leisure activities.

Dr Waitkins tells the harrowing tale of a thirteen-year-old boy who would have benefitted from a greater awareness of the lethal consequences of contact with rats. The boy's hobbies included keeping a rabbit in a hutch at the bottom of his garden in Oxfordshire. One day, while attempting to clean out the cage, he came across a rat which had moved in and snuggled up next to the rabbit to keep warm. The startled rat, frightened by the boy's attempt to push him out of the hutch, immediately urinated over the child's hand and jumped out of the cage, escaping into the garden. The boy continued to clean out the hutch. Within ten days, he was dead.

There are many similar cases in which the victims failed to see their symptoms as those of Weil's disease, unaware of any link between their illnesses and any recent exposure through work or leisure activities to contaminated water. In 1987 a man from Essex caught the disease from rat-infected water at a leisure centre and died. In the same year, another Essex man died after catching the disease at work – he was a farm stockman. In September 1988, a water skier in Warwickshire died four weeks after swallowing contaminated water.

One particular case in which leptospirosis was diagnosed in time ended up in court. In November 1988, Mr Justice Potter awarded £125,000 to a construction worker against his employer after the worker had contracted leptospirosis. Mr George Campbell had to be put on a life support machine having contracted Weil's disease while working on a canal bridge. Mr Campbell had cut his hand while moving equipment which had been stored on the banks of the New River Canal in Finsbury Park, north London, where rats were known to be present. The defendant claimed that his employer should have taken steps to kill the rats on the site.

Of all occupations today, it is farming which carries the greatest risk of exposure to leptospirosis. Weil's disease is now accepted as an occupational hazard of farm work, and is recognized by the courts as an industrial disease for compensation purposes. Between 1933 and 1946 there were 45 recorded cases of leptospirosis amongst farm workers but between 1978 and 1986 that figure had shot up to 285 cases. In a group of farm workers studied by the Leptospira Reference Unit last year, 11 % were found to be carrying antibodies which showed that they had been exposed to the disease in the previous two years.

Water sports have become increasingly popular around Great Britain over the last few years, with an estimated seven million people participating in rowing, windsurfing, water skiing and canoeing. In the same LRU study, an alarming 2 % of people regularly taking part in water sports were also found to be carrying leptospira antibodies. With over one million people canoeing up and down the country, it is a fact which has not gone unnoticed by the British Canoe Union, based in Nottingham. Geoff Good, the BCU's Director of Coaching and Coordination, takes the problem of leptospirosis very seriously.

Every member of the British Canoe Union is told about Weil's disease. They each receive information about the nature and symptoms of the disease on the back of their handbook, and a pink card which they are urged to keep with their membership card.

The 15,000 members of the BCU are warned of the dangers of Weil's disease by Dr John Whitehead, who sits on the BCU Medical Advisory Panel. Dr Whitehead's message on the pink card is clearly stated:

Weil's disease can cause serious illness or death. The risk of infection is greater where stagnant or slow moving water is involved, but cases have occurred on swift moving streams, as well as lowland rivers. The bacteria are absorbed through the skin and mucous membranes of the mouth and eyes. It gets into the blood stream more easily if you have a minor cut on your skin or feet, or if you do capsize drill or rolling.

Canoeists are given specific advice: avoid capsize drill and rolling in stagnant water and always shower after canoeing. Water-proof plaster is recommended to cover cuts and scratches, and the use of foot-wear is advised to avoid cutting the feet. Most of the emphasis is placed on the early identification of the illness, and canoeists are told to remind their doctors of the Leptospira Reference Unit in Hereford or to ask for an ELISA test at their local public health laboratory. 'Tell your doctor you have been canoeing, and ask if you can have a blood test for Weil's disease,' it urges, in capital letters.

Geoff Good believes that public education is one answer to the problem, and is teaming up with a colleague from the Royal Yachting Association to plan a one-day seminar with the Central Council of Physical Recreation for all water sports associations. He hopes to invite lecturers who are authorities on leptospirosis and other forms of water pollution, to inform people of the risks they face when they are participating in water sports.

In February 1989, the British Canoe Union began an active campaign aimed at lowering the incidence of Weil's disease among its membership by writing to various bodies and local authorities seeking positive action on three fronts.

1)   To ensure that more GPs are aware of the occurrence of Weil's disease among water sports enthusiasts. The BCU Medical Advisory Panel wrote to the British Medical Association to alert doctors to the possibilities of the disease. The BCU felt that the BMA had paid insufficient attention to the disease in the past. There was evidence, it claimed, that not all GPs had been alerted, and in some cases, doctors failed to take prompt action resulting in severe illnesses which could have been avoided.

2)   To ensure that sufficient testing apparatus was available to local laboratories for immediate ELISA tests to be carried out.

3)   To seek positive action with regard to the extermination of rats, with particular reference to the recent increase which the BCU estimated at a national average of 70 %.

The BCU's policy of education and information contrasts sharply with that of the National Federation of Anglers. In a poorly presented reissued leaflet sent to its 400 club secretaries, the Federation warns of the dangers of leptospirosis in just half a page – and offers little detail on what to do if one of its 280,000 members suspect having contracted the disease. Frank Bullock, the Federation's administration officer, says that he is aware of the dangers of Weil's disease but has no plans to issue cards or more information.

Geoff Good makes a strong case for financial investment in research into testing and treatments for leptospirosis, suggesting more money be found for the Leptospira Reference Unit. 'It does a fantastic job,' he says. 'This is a disease which affects people involved in healthy recreational activity – it is a risk which shouldn't really exist. With more money, the Unit could help to alleviate the problem.' Mr Good's long term conclusion is simple: exterminate the root cause of the problem. 'Without rats there would be no Weil's disease; that is the real answer to the problem of leptospirosis, getting rid of the rats. The increase in the rat population is obviously a great concern to the BCU, because statistically, the chances of coming into contact with Weil's disease is likely to increase.'

Dr Waitkins agrees. Her Unit in Hereford sees over a thousand positive test results each year, reflecting between 100 and 150 patients, and the number is growing. 'The only way to completely control Weil's disease is to eliminate the rat, which to me is an almost impossible task,' she says. 'The answer is to understand the source of the problem – rats.' Dr Waitkins believes detailed research is needed into Britain's rat population, from size and habits to patterns of disease. 'We have to find out if they are coming into contact with more humans, and exactly how great that risk is – and perhaps look for things which can be put into the environment.'

Of course, attempts to find a vaccine are being urgently researched around the world, but it is a slow process as there are 207 different varieties of leptospirosis. If someone in Britain were to contract the *ictero-haemorrhagia* type, they would still be able to catch another strain abroad, with no cross-protection. Although there are dozens of people working on vaccines internationally, all using different approaches, Dr Waitkins is pessimistic about their chances of short-term success. 'My scientific projection is that I don't think we're going to have a vaccine which is going to be of any use for the next 10 to 15 years.'

According to the World Health Organization, deaths from leptospirosis are far worse abroad. In China, for example, where no adequate rodent controls exist and paddy fields provide ideal living conditions for the bacterium, a recent outbreak resulted in nearly 4,000 cases, with a 50% death rate. In the humid tropics of Barbados, where a health organization conducted a study, people in hospitals suffering from fevers of unknown origin were all tested for leptospirosis, and in one sample 50% were found to have the disease. Nearer to home, in Italy last year a rat got caught in the water supply of a large city, resulting in 89 cases of Weil's disease and four deaths.

Dr Sheena Waitkins is clearly frustrated at the lack of action taken to keep Britain's rat population in check. When the Unit first arrived in Hereford in 1981 it was sent 668 sera to test. 1988's figure is over 6,000 and she claims the Unit could not cope if inundated with primary testing. 'Leptospirosis is a preventable disease. That's the tragedy. You can prevent deaths occurring by early treatment. But in a country like Great Britain a disease like leptospirosis should be properly investigated. We do have a problem, and it's one that will grow.'

With cases of leptospirosis already rising in the UK, the new effect of a sudden and dramatic growth in the rat population is bound to increase the chances of catching Weil's disease. 'Just imagine rats as little bags of infection,' says Dr Waitkins, 'the more bags of infection there are, the more likely you are to bump into them.'

7

Bags of Infection

Rats are almost literally 'bags of infection' carrying, as well as leptospirosis, a myriad of different diseases. Unlike many other mammals which are well known for carrying one specific form of disease, such as foxes, which host the virus that causes rabies, rats carry several organisms which can cause serious illness or death to humans. A detailed examination on every known disease carried by brown rats could easily fill a book by itself. This chapter will briefly discuss the better known diseases which rats pass onto man and look at methods of rodent control.

The best known and most feared rat-borne disease in the history of mankind is without doubt the bubonic plague. The main culprit, as we have seen, was the black rat – *Rattus rattus*, but it is important to note that the brown rat is equally capable of carrying the disease. Although the most recent cases in the UK date back some eighty years to around 1910, the plague is still with us. In 1980, there were 142 cases of plague reported in the Americas and the worldwide figure numbered 505. Western Europe is today considered plague free, but there are many countries in the world, including parts of western USA, which are still regarded as plague areas.

A disease or infection which is transmitted from a vertebrate animal like the rat to man is called a zoonose. Of all the viral zoonoses, rabies – often known as hydrophobia after its symptom of an irrational fear of water – is one of the most feared. It is normally fatal in man, and can be transmitted by the bite of a warm-blooded animal. The disease is most commonly associated with a dog bite. Experts have been unable to clearly link rodents with transmitting rabies to man, but throughout the world it is still common practice to give rabies vaccinations to patients suffering from rat bites.

There are several other viral diseases which rats carry, which are passed on to man through tick bites; fortunately these are not known in the UK. Two of the strangest sounding of these are Russian Spring-Summer Encephalitis and Central European Encephalitis. Rats still carry Lassa Fever, mainly in the African continent. It is passed on to humans through food contaminated with rat urine and is widespread in some parts of Africa, where the human fatality rate can be as high as 50%. Hantaan Fever is a newly-discovered disease in Europe and the USA, although it has been known for a long time in the Far East. The disease is reported to have been spreading over the last ten years, but has not so far been detected in Great Britain.

One topical bacterial disease, salmonellosis – food poisoning – is on the increase throughout the world, and especially in Britain. There are over a thousand different species of salmonella and some rats are thought to carry three types. Rats often live in areas where food is stored, and it is easy for them to contaminate grain, rice and wheat with urine or droppings. Although there are thousands of reported cases of food poisoning in the UK each year, with many more going unreported, most of them are attributed to improper food handling, storage and processing. Nevertheless, there are a growing number of cases of salmonella poisoning which have been directly traced back to rat contaminated food.

Of the three types of typhus – epidemic, murine and scrub – rats have been linked to the last two. As with the plague, the disease is spread by the flea *Xenopsylla cheopsis*. The death rate amongst man is around 1% of those infected even though the most heavily infected rats do not die from the disease. Scrub typhus is confined to the Far East – Japan, Asia and Australia. It is also known by its Japanese name, *tsutsugamushi fever*, after its causative organism, *R. tsutsugamushi*. It is transmitted to man by the bite of infected mites which live on the rat – the mites are often known as 'chiggers'. Another rare and strange sounding disease, mainly found in the Americas and especially in the USA, is Rocky Mountain spotted fever. Again, it is transmitted to man through the bites of ticks, usually the Rocky Mountain wood tick. Although the fever can be carried by rats, it is more commonly spread by dogs. In 1977 there were more than 1,000 cases reported in the USA.

Rats also play a part in the spread of Chaga's disease, toxoplasmosis, trichinosis and pneumocystis carinii, as well as carrying the liver worm, *Capillaria hepatica*, and several fungal diseases. Another bacterial disease is rat bite fever, but the chances of contracting it from a rat bite are very small indeed. If left untreated, rat bite fever has a mortality rate of around 10%. Rat bites themselves are a rare occurrence in the UK, but they still prove a problem worldwide. In America, conservative estimates put the number of rat bite victims at an average of 14,000 each year. New York City alone averages several hundred per year, and in Illinois the figure is over 2,000.

Rat bites are more common in the deprived areas of cities. The main victims are under the age of 12 and it is generally hands which are bitten. A suggested reason for this is that children try to pick rats up, although some authorities argue that rats attempt, especially in the cases of small children and babies, to take food from hands and faces. The bites are usually deep slices caused by the rat's razor sharp teeth and they bleed heavily. A rat bite rarely results in death, but there are obvious risks of infection from tetanus, leptospirosis and rat bite fever.

Despite popular belief, rats do not often bite or attack people, unless cornered or threatened. It is also claimed that when trapped, rats 'go for the throat'. In most cases, rats run for the lightest area they can see, often mistaking the exposed white throat for an exit.

Amongst the other bacterial diseases, yersiniosis, caused by organisms related to those which cause plague, is found in rats. It is passed onto humans through the contamination of animal foodstuffs with infected urine and droppings. Animals which then eat the bacteria pass it onto man through contact. The same is true of erysipelis, which usually affects humans who come into regular contact with animal carcasses.

With the exception of leptospirosis, rat-borne diseases are not a major problem in the UK today. The majority of the more dangerous ones are contained and controlled throughout the world. Although people still fall victim to the plague, Lassa fever, typhus and rabies, their numbers pale into insignificance when compared with the millions of plague victims throughout history.

However, most experts agree that controlling the rat population is not just a fire-fighting operation, to be done when the odd plague or epidemic arises. It should be a regular commitment, aimed at constantly reducing the size of the rat population. It only takes a few rats sneaking aboard an aircraft from Asia or Africa, or smuggled in by accident in a food container on a lorry from Europe – or indeed scurrying along the newly opened Channel Tunnel – and the British rat population could be infected with diseases hitherto unseen in the UK.

Rats will always pose a danger to our health. They share the same food as humans and live in close proximity to our food stores and shops. The potential risk is ever present, but with containerization on the increase and imports at an all time high, that risk is increasing.

It is ironic that while the UK's rat population continues to grow techniques for rat control have never been better. We have come a long way from His British Majesty's Ratcatcher of the 18th century and the organized rat fighting of Victorian Britain. The Lord Chamberlain's office records the last official rat-catcher in Britain was employed in 1908. He still used a combination of poisons and traps but didn't wear a splendid uniform like old Jack Black. Today's twentieth century breed of ratcatchers wear white coats and they have perfected their trade to a science.

Before the Second World War, ideas for killing rats in large numbers were still wild and fanciful. In Moscow, for example, in the 1930s, a group of scientists hit upon the idea of breeding a race of super rats. The theory was simple. Starve the rodents of food and water until only the strongest remained alive. Breed these survivors with each other, and then repeat the process. The master plan was to release the super rats into the towns and countryside, hoping that they would in turn eat the weaker ones, and thus through a form of cannibalism reduce the size of the population. The American Department of Agriculture at the time wished them all the luck in the world, but told them not to be too disappointed if they were to find they had bred a race of super chicken-killers and baby-biters!

Up to the middle of the 1940s, new methods of baiting were researched, and experiments were carried out to find the most efficient poison. As we have seen, rats fear new objects, and getting them to accept poisoned bait took some skill. The newest method involved giving them bait which had not been poisoned, then gradually topping it up each day, until they became used to it, at which point poison was added. Arsenic, strychnine and phosphorus were used, all of which proved dangerous to other animals as well as to humans. Phosphorus was particularly nasty in that it literally burned away the insides of the rat.

1949 marked the breakthrough in rat control, with the advent of Compound 42. In the United States some twenty years before, there was a widespread outbreak of a cattle disease known as 'sweet clover disease', which brought on fatal haemorrhaging among cows. The cause was found to be a product of spoilt clover hay, called dicoumarol. By 1949, when scientists realized that it made an efficient rat poison, a purer substance was derived from dicoumarol and sold under the name of warfarin. Within 15 months from July 1950, over one million pounds of the concentrate were sold under 300 different names throughout the world. Over 100 articles were written about warfarin; at last, they proclaimed, a safe and effective rat poison had been found.

Warfarin is an anti-coagulant, which means it upsets the blood's normal clotting mechanism. Blood clots usually form to prevent the excessive loss of blood after damage to tissue, internally or externally, and to aid the body's healing process. In the case of rats, anti-coagulants not only prevent clots from occurring, but cause deprivation of oxygen to the brain. Rats take the bait, which is very low dosage, over a few days. The rodents get slowly weaker, and generally die within five days.

Warfarin is much safer for non-target species, such as domestic pets and wild animals. Unlike arsenic, the concentration is low, and it is therefore safer on animals which eat rats, too. By the middle of the 1950s, warfarin was the world's most popular rat poison. But it was a short-lived victory. By the beginning of the 1960s, reports began to filter through of a generation of rats which were resistant to warfarin. Super Rat had arrived.

Resistance is defined as the ability of an organism to survive the effects of concentrations of a chemical. In the case of the UK's rat population, a small number of rodents were not killed by the standard dosage of poison given to some colonies. This nucleus of rats had a greater ability to survive; they interbred and the result was a second generation of rats which inherited the ability to withstand the same poison. This small group soon became the majority in some colonies, and they started to spread around the country.

The first cases of super rats resistant to warfarin were discovered in Scotland in 1958 and Wales in 1960. It was also discovered that black rats, although a minor pest in the UK, were equally hard to control once resistance had taken hold. Small colonies of resistant brown rats sprang up all over the UK in the early 1960s, the worst outbreaks occurring on the Anglo-Welsh border.

The rodent control industry was not prepared for this setback. It had become complacent, relying almost exclusively on warfarin, and had almost shelved its search for any possible alternatives. Older poisons were once again called back into use while the scientists donned their white coats and took to their test tubes in search of a new generation of anti-coagulants.

Parliament made it harder for the research chemists in 1963 with a legal restriction on the extermination of rats. The

Prevention of Cruelty to Animals legislation effectively banned the use of the older and more dangerous treatments – outlawing arsenic, strychnine and phosphorus poisons and encouraging the development of more powerful anti- coagulants.

Peter Bateman, the Director of Public Affairs at Rentokil, Britain's biggest pest control business, is guardedly convinced that the humane laws are kinder to rodents.

> We were told that you had to kill rats in the most humane manner possible – without causing them any symptoms of pain or suffering. So we banned phosphorus and instead gave them anti-coagulants, which caused liver anaemia, so they died slowly. They still have to be killed with kindness, in a humane way, according to the Control of Pesticides Regulations. This accords with our policy, although the rat is the unacceptable face of the environment.

In 1974 the pest control industry fought back. A major breakthrough was announced, heralding a second generation of successful anti-coagulants: difenacoum, brodifacoum and bromadiolone. It is largely this generation of poisons which are in use today. Although science had once again triumphed over the rat, the industry had learned its lesson and today millions of pounds are spent each year on monitoring the effects of current rat poisons and developing more effective baits and treatments.

There are, of course, other methods of destroying rats which are still legal in Britain today. Liquid baits which rats drink, gel baits which they rub up against and subsequently lick off their fur and paws and gassing with aluminium phosphide which releases phosphine gas are only used when regular food baiting proves difficult. Mechanical traps like the 'back breaker' are still permitted by law. As its name implies, this trap closes sharply on the rodent's back, killing it immediately. But worldwide, it is still warfarin that is the most popular of all rodenticides. Difenacoum and brodifacoum are only used when warfarin-resistant breeds of rats are discovered.

The technical battle, at least, against rodents has been won. Today we are armed with the soldiers and the weapons in our war against the rat. We understand its patterns of feeding and breeding and have developed successful methods of treatment and extermination. We know which poisons are the most effective without causing us or other creatures harm. Why then, with our armoury of scientific knowledge and specific chemicals are we currently suffering from an acute increase in rat infestations all over the country? To answer that question, we must first find out who exactly is responsible for the public control of rats in Britain.

Part Three

THE NEW PLAGUE

8

Rats on the Rates

The Prevention of Damage by Pests Act 1949, which super-seded the Rats & Mice (Destruction) Act 1919, came into force on 31 March 1950. It made local government the front line in the war against rats. The Act of Parliament, which is still in force today, placed the responsibility for rodent control firmly in the hands of the local authorities, under the watchful eye of the Ministry of Agriculture, Fisheries and Food.

The purpose of the legislation, which was applied throughout the country, was 'to make permanent provision for preventing loss of food by infestation'. National action over the rat problem was to be taken on a local level. Local authorities were charged with 'the duty to take such steps as were required to ensure that their districts were kept free from rats and mice'. They were to make regular inspections, destroy infestations wherever they found them on their own land, and enforce those duties on other landowners and occupiers. Local authorities were given complete powers of entry for the purpose and legal guarantees for the recovery of their expenses. In the Act, owners and occupiers of property other than agricultural land were obliged to give notice in writing if they discovered 'substantial numbers' of rats.

The Act is now some forty years old and its critics claim that it is due for an urgent review. They argue that expressions like 'substantial numbers' are too general and need clarification. Calls for the removal of the derogation from agricultural land in requiring formal notice to be given to local authorities that rats are on the property have been heard. But despite its age and the need for amendments, the Prevention of Damage by Pests Act remains the single most important piece of legislation enacted on rodent control this century. 'Rats on the rates' became the order of the day as local environmental health departments took charge of its administration. If the public were obliged to be foot soldiers on the look out for signs of infestation then the environmental health officers became the sergeants and the officials at the Ministry the generals.

During the thirty years from the 1950s to the 1970s, the Ministry for Agriculture, Fisheries and Food (MAFF) played a central co-ordinating role in supporting research and development in rodent control. Pest Control Liaison Committees were established which organized speakers and provided information to local authorities. National surveys were undertaken, providing centralized information on the level of rat infestations around the UK. At the end of the 1970s however, in one of the many local government reviews of the period, the decision was taken that central government should no longer interfere in the local authority business of rodent control. MAFF reduced its general training programme and disbanded its committees, although it still provided speakers for some authorities. Just as important was the halting of MAFF's extensive research programme into pest control, with the exception of its investigations into warfarin-resistant rats.

MAFF's work on the development of methods of urban rodent control and its national survey programme both came to an end. Although MAFF still operates a rodent information service, offering its relatively expensive treatments to farmers and local authorities, as well as running courses at colleges, it is a shadow of its former self. MAFF's function as a unique central co-ordinating body in rodent control has not been replaced; there is no Government body actively involved in national rodent control. It is the absence of central information gathering that has made effective planning and policy-making on a national scale impossible.

Without a countrywide survey of rat infestations in the UK, environmental health officers have been left to their own devices on the collation of statistics. The efficiency of each pest control operation varies from authority to authority; some councils keep no records at all while others plan extensive rat baiting programmes and chart their progress. It is for this reason that the recent increase in the size of the UK's rat population was not spotted sooner. EHOs who had been faced with a rise in reported infestations and had to provide more baiting treatments could only see their local problems. They had no real way of knowing what was happening around the rest of the country. Although EHOs knew that, within their own authorities, rats were on the increase, there were no real comparative national figures.

In February 1989, the Association of Metropolitan Authorities published the results of its survey 'Rats on the Rise', taken amongst its member authorities in Greater London, Merseyside, Greater Manchester, West Yorkshire, South Yorkshire, Tyne and Wear and the West Midlands.

The AMA's report, which was based on returns from 18 authorities, followed the increase in rat infestations over the previous year, and produced some worrying statistics. Rodent complaints had doubled in Wakefield, risen 45% in Harrow and 28% in Barnet, and over two years had risen 50% in Newcastle. The report noted that '. . . infestation levels would appear to be rising in most of the metropolitan areas, particularly where shortages of funds are evident'.

'Rats on the Rise' drew attention to another vital factor in the equation – the role of the water authorities in rodent control. It is the responsibility of the water authorities, as it is with every other public or private body, to keep their land, in this case their sewers, as free from rats as it is able. In practice, it is impossible to bait rats to extinction in the sewers, but water authorities are obliged to keep the infestations down to a minimum level. Most local authorities carry out sewer baiting on behalf of water authorities who pay them a negotiated sum for the contract. But according to the AMA survey, of the few local authorities that had been questioned, all complained about the level of funding that they received from their water boards.

The chairman of the AMA's Environmental Health and Consumer Protection Committee wrote, 'It is clear that water authorities are cutting their budgets for rat control in the run-up to privatization – and things are bound to get worse once they are only in business to make profit for their shareholders.'

Claim and counter claim flew. The survey, according to the water authorities, was a political move prompted by political opposition to the government's programme of privatization. Their argument ran that, had the government not decided to sell off the water authorities, no such survey would have been undertaken and the rat problem would have been dealt with in much the same way as it had been for the last ten years. The AMA on the other hand described the matter as a financial issue with health implications; cutbacks in water authority spending were taking place all over the country and more rats threatened the country's health. The AMA announced that its survey had in fact been prompted by complaints from Birmingham's authority that the Severn Trent Water Authority, despite estimated profits of around £160 million for 1987, had refused to provide an extra £150,000 for more sewer baiting teams at a time when its rat population had increased. It was more rats, it claimed, rather than politics which had brought the situation to a head.

Local authorities around the country were facing a significant increase in the size of the rat population and were having to contend with two serious handicaps: pressure on their rat-baiting and sewer maintenance programmes in the form of financial restraints imposed by central government and water authority contracts which had not kept pace with inflation. In many cases, the water boards' allowances to the local authorities had remained the same for three years.

Over the last twelve months, dozens of 'Shock! Horror!' stories on rats appeared in the pages of the national press. People became more aware that there was a problem as journalists reported on the millions of rats which were finding their way out of the sewers through crumbling drains and broken tunnels. Several incidents of rats biting children were covered. The Chief Environmental Health Officer of Bristol complained that a reservoir of rats had been crawling out of the Victorian-built sewers and finding its way into homes throughout the city. In Birmingham in September 1988, an average of five complaints per week from restaurants with rat infestations were reported; more than half of the city's sewers were infested with rats. In London, a report by engineers showed more than half of its sewers were similarly affected. Anyone keeping a press cuttings file on rat-related stories in the newspapers would have seen it double in size during 1989.

The AMA survey, however, did not make the headlines; at most, it achieved its aim of getting the debate on the rat increase aired. It was not a complete national survey – 18 replies from a total of 62 member authorities did not present a complete picture. A national survey needed to be just that; it had to represent at least half of its member authorities. It was the turn of the Institution of Environmental Health Officers, an independent non-governmental body of 7,500 strong, representing 90% of EHOs in England, Northern Ireland and Wales (Scotland has its own body). The IEHO is a non-political body established by royal charter and reflecting, according to its charter, 'an independence of view, of judgement and of practice with environmental health as its sole motivation'.

The Institution carried out a detailed survey amongst its members to determine a true picture of the extent of the rat problem around the UK ('Rats: A Survey of Incidence of Rats in England and Wales, 1988–89'). Questionnaires were sent out in March and the results published in August. Of the 402 local authorities in the UK excluding Scotland, 244 replied, giving an overall response rate of 61%. The Institution's findings proved beyond doubt that a national rat problem existed and pointed the finger at several factors which had caused it.

The total number of premises treated for rat infestations rose sharply from 155,000 for the period 1987–8 to 210,000 for 1988–9, an increase of 20% in just one year. This figure represents the average increase in the size of the rat population in the UK. It is further corroborated by the increase in the use of rodenticides by local authorities – up 22% – and by Rentokil's rodenticide sales figures for the same period – up by around 20%. Over 78% of local authorities were carrying out regular sewer baiting programmes, but three quarters of them, a significant majority, had not had their funding increased in response to the growing scale of the problem. 30% of them had actually had their funds reduced in real terms.

It is interesting to note that the IEHO survey was unable to come up with a figure in answer to the simple question, 'Level of funding received from water authorities to carry out sewer baiting programmes during 1988'. The writers of the report comment that the funding formula differed up and down the country, and it would be misleading to simply total their findings. However, they did note in the case of Thames Water Authority that the specific figure spent on sewer baiting in their area was not itemized in its annual report, and could not be provided to them. The IEHO believes this case to be typical of most water authorities, and warns that water authorities' investment level in sewer baiting in real terms has been reduced nationwide.

The survey highlights several reasons for the increase in the number of rat infestations. The winter of 1987 produced above average temperatures and the winter of 1988 was the second warmest winter since records began in 1659, according to the Meteorological Office.

As we have seen, mild winters coupled with above average rainfall levels provide ideal living and breeding conditions for rats. Wet summers have also played their part: pregnant females have less far to travel for water, there is an increase in crops and fruit and often flooded underground waterways release hundreds of rats to the surface. But the report notes that although the mild weather is an important factor in explaining the increase, it is a combination of other, preventable elements which has made the problem significantly worse.

The water authorities came in for another battering. The

inadequacies of the different authorities 'were the most frequently mentioned causal factor with regard to the rat problem.' The report cites one example of a reply from a local authority:

A succession of mild winters has contributed overall, with the increase in the city centre being exacerbated by Southern Water cutting sewer treatment funding to a third of last year's level.

Reduced funding is a popular complaint – the IEHO report quotes another local authority's comment in full:

We share the fear voiced by other local authorities that the new water authorities will pare to the bone sewer treatments if they are given the chance. Already our Regional Water Authority has asked us if we really need to carry out our present methods of baiting, rather than doing something less intensive.

The water authorities were further criticized for having no coherent programme on regular sewer baiting, and concern was expressed that they followed a policy of cure, rather than prevention, and that they were slow to co-operate on remedial work.

It is one-sided to suggest that the current rat problem is solely the fault of the water authorities. The survey highlights other areas of concern, especially a general lack of funds available for local authorities to properly do the job.

We often hear a cry from the public sector that financial cuts are responsible for most local government ills. Politicians from all sides of the House lambast us with figures: adjusted figures, real-terms figures, inflation-related figures and comparative figures, all seeking to prove their political arguments. There is no doubt that local government is undergoing a change. The present Government, we are told, wants to see it operate on a more commercially sound footing. Local authorities are being encouraged to put some of the services that they provide up for tender and to become in effect agencies, rather than contractors. As part of this process, many local authorities have found their budgets greatly reduced. As we shall see when we examine the work of Rodent Control Operators in Manchester and London, the current cuts in local authority budgets from central Government have had a radical effect on the problem of rat control.

Most commonly, in the IEHO's survey, financial shortages were blamed as the main source of trouble. The increase in complaint levels has not been matched by staff or resources, and many Pest Control Departments can now only afford to offer a reactive service. A lack of repairs to older sewage and drainage systems has resulted in conditions ideal for rats; the capital expenditure necessary for their repair is not available. Some authorities have been forced to charge for refuse collection from commercial premises, resulting in more rubbish on the streets as private customers attempt to avoid expense. The increase in demand for these reactive treatments has led to less staff being available for regular treatment programmes.

Changes in agricultural practices were blamed by most rural authorities. The agriculture industry was criticized for:

1) Cutting their expenditure on pest control.
2) Failing to notify councils of rat infestations.
3) Poor proofing and storing methods.
4) The change from dairy farming to grain production.

Farmers were also particularly criticized for the indiscriminate use of warfarin, contributing to the development of resistance to that poison among rats.

Finally, the survey touched on other general factors responsible for the increase in the rat population. British Rail, in common with other public bodies, was reported to have cut back on its rodent control operations. New methods and materials used in building developments such as drains not properly sealed and plastic replacing metal have caused building sites to become new places of harbourage and have all contributed to the overall picture. The report also notes that Britain's streets have never been dirtier, with the proliferation of litter particularly from fast food ships. It is the food waste which has created so many opportunities for rats in terms of feeding and harbourage. Last but not least fly tipping also contributed in attracting rats to an area. The IEHO's many recommendations will be discussed further on in the book when we look at the possible solutions to Britain's rat problem.

It is not always easy to fully understand a problem when it is discussed in strictly statistical terms. The sentence '30% of local authorities have had their funding reduced in real terms this year' tells us nothing of what that actually means to the ratcatchers and pest control officers whose daily job it is to exterminate rats. In the following chapters, we take a detailed look at two very different local authorities – Manchester City Council and the London Borough of Barnet. We will examine the effects of the increase in the rat population on the people who are propping up the dam, asking them who or what is reponsible, and then look at the water authorities and how their funding policies are affecting rodent control.

9

London's Rats

David Drake is a Rodent Control Operator working for the London Borough of Barnet. He calls himself a ratcatcher and enjoys his job, which he has been doing for the last three years. He used to be a butcher, but when the opportunity arose to change direction, he jumped at the chance. Drake explains that he likes meeting the public and providing a good service. He is especially fond of his current position because, he says, 'People are always so pleased to see you!' His workload, however, has increased over the last two years as the number of call-outs has doubled. Just one afternoon with David Drake and his immediate boss Mark Jervis, Barnet's Pest Control Supervisor, is enough to convince anyone that the borough's rat problems are increasing at an alarming rate and that there are not enough staff and resources to cope.

Our first stop is a colony of rats reported to be living behind a shopping parade near Burnt Oak underground station. Driving through the area, down the Edgware Road and past dozens of shops, warehouses, discount DIY stores and car showrooms, it looks much the same as any other average suburban district of London. There are no large mounds of black rubbish bags on the pavement and no more litter than on any other main road and no piles of boxes full of rotting food or beer cans. Turning off and heading up a hill takes the driver towards a residential area with fairly wide roads, plenty of trees and the occasional grassy verge with rows of ordinary looking houses.

But a couple of left and right hand turns past the station and a small parade of shops reveal a totally different environment. The area looks like something specially created for a shoot-out scene in an American cops and robbers movie, set in the slums of downtown Manhattan or Brooklyn. We are standing at the rear of a parade of high street shops facing a small waste area. It is a quarter of the size of a football pitch, surrounded on three sides by buildings, backing onto the trees and bushes of an embankment, divided by a wire fence. There is litter all over the ground, cans and bottles of every kind, packaging and boxes and a surplus of food waste. Raw meat leftovers are scattered here and there. The shopping parade building is on two floors. One of the shops is occupied by a butcher and there are five sides of meat lying on a cement wall directly outside its door on the first floor. Another shop is a kebab house; evidence of this can be seen by the dozens of meat packaging cases and wrappings strewn over the ground floor passage and around the waste area. It is a blisteringly hot day, and the smell is foul and flies are everywhere, particularly hovering over the scraps of entrails and offal falling out of torn rubbish bags, lying at the base of the dustbins. Looking up, we see someone washing down the balcony at the back door of the butchers. The drain is blocked, and from our position we can see that a combination of grease and meat bits is preventing the water from passing through the pipes. Soapy waves overflow and spill directly onto the ground, forming dirty puddles of grease on the muddy rubbish below.

'This place is a Garden of Eden for rats,' says Mark Jervis, pointing to the solidified fat down the side of the wall. 'We try and close them [the offending shops] down, but they wait until the very last minute – it often takes months for the law courts to hear the cases – then they offer to pay a fine and fix things up. But within a month, they do it again. We don't have the powers and the fines are unrealistic.'

David Drake demonstrates graphically the three elements essential to a rat's survival: 'Over there's the water, the food and the harbourage . . . and there is your rat colony.' Experienced eyes look beneath our feet, 'Oh! I'm standing on a hole here,

can you see it?' The embankment on which we stand is 'live';
full of rat holes that are still in use.

The ratcatcher explains that there are four ways to spot signs
of rat habitation. The first is looking for a rat run – a smooth
strip of ground, usually on mud, which is worn down by the
constant pitter-patter of tiny paws. Secondly, searching for
grease marks on walls or posts. Rats living in the dirt and grime
of sewers or in muddy burrows are covered with filth and leave
greasy smears on objects they brush up against. Looking for tail
or foot markings in the dust is another way of finding rats.
Finally rat droppings are a clear indication. Our waste site
boasts a myriad of signs – rat sitings were reported previously
by the residents of the flats opposite. On that occasion David
Drake found the colony and left bait around the area. Today he
has returned to examine the takes, and decide on the treatment.

Plastic bags are filled with poisoned bait and left at the entry
points of the colony and around its runs. A good take is when
most of the contents of the bag have been eaten. It is not a job for
the squeamish, and David admits that he has often been
surprised by a rat jumping onto his feet or over his hands.
'I wear gloves a lot of the time,' he says, 'but it still gives me a
start.'

While David checks his takes, Mark Jervis demonstrates the
detective-like quality required of his work. One of the rat holes
has not been used – there are flowers at the mouth of the hole
and the flowers' seed heads and stamens have not been touched.
Had the hole been an active one, the flowers would have shown
signs of being jostled. 'You learn these things on the job,' says
Mark, 'for example, it's easy to tell the difference between an
entry hole and an exit hole. The exit holes have no spoil around
them, unlike the entry holes, which have mounds of mud where
the rat's been digging!'

Mark Jervis explains that the waste ground is the responsibility of the shop owners who will be told to clean up the mess
and maintain proper standards of hygiene. The slope of the
embankment, however, where the main burrows are located, is
situated on wasteland owned by London Regional Transport,

Overflowing rubbish bins are a major food source for rats, often tempting them away from poisoned bait.

Sewer rats can carry myriad diseases on their fur and paws, passing them on to humans through contact with contaminated surfaces.

Rattus rattus – the black rat – is especially good at climbing. It is slighter and more agile than its brown counterpart.

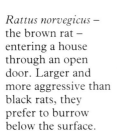

Rattus norvegicus – the brown rat – entering a house through an open door. Larger and more aggressive than black rats, they prefer to burrow below the surface.

It has been estimated that a healthy pair of adult rats in perfect conditions can be responsible for nearly 1,000 descendants in one year.

Rats can reproduce when only three months old. The gestation period lasts for twenty-one days resulting in an average litter of eleven pups.

Not everybody hates rats. In Deshnoke, India, a man sits in the grounds of a 'rat temple' where visitors are encouraged to feed them.

Rat damage to an electric plug. Brown rats gnaw at a pressure of 500 kg per cm², producing six bites per second.

Rat damage to a lead water pipe. The lower incisors of the brown rat have a hardness of 5.5 on Moh's scale, compared with steel which measures just 4.

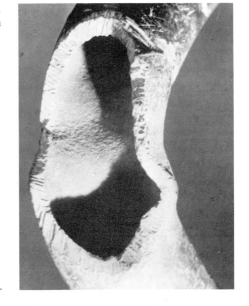

Around 60 per cent of British rats are infected with leptospirosis which they carry in their kidneys. Rats excrete the bacterium through their urine often into water where it can live for up to forty-five days.

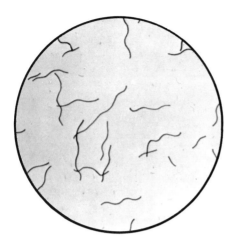

Under the microscope – the deadly leptospires. Lepto is Greek for worm and spiro means coil – an exact description of this highly motive organism.

A Weil's disease warning card issued to all members by the British Canoe Union as part of its information campaign.

WEIL'S DISEASE

RIVERS, PONDS AND CANALS ARE USUALLY INFECTED WITH A BACTERIUM WHICH CAN CAUSE WEIL'S DISEASE. MOST OF THE TIME WEIL'S DISEASE MAY TAKE THE FORM OF A CHILL OR POSSIBLY RESEMBLE AN ATTACK OF FLU.

WEIL'S DISEASE CAN CAUSE SERIOUS ILLNESS OR DEATH

YOUR DOCTOR IS
REMINDED OF THE EXISTENCE OF:

**The Lepto Spirosis Reference Unit
Public Health Laboratory
County Hospital
HEREFORD. HR1 2ER**

Telephone No available from BCU 0602 691944

Results of blood tests have been known to take two weeks or longer through the normal laboratory system. Deaths and serious illnesses have occurred because of slow identification.
Your local laboratory can provide your medical practitioner with a result within 2-3 hours through an ELISA test

KEEP THIS SAFELY WITH YOUR MEMBERSHIP CARD

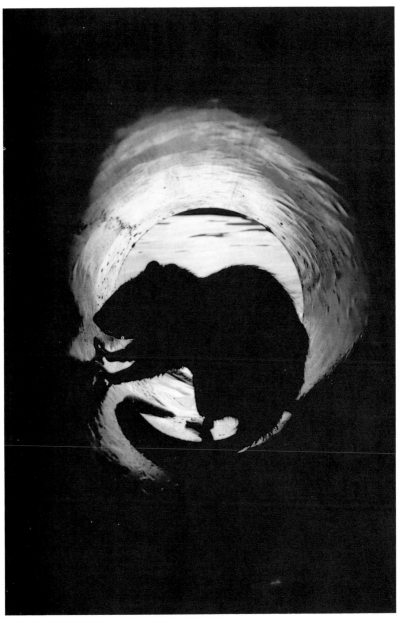

The majority of Britain's rats live below ground. Yet despite our technological advances, we still seem unable to get to grips with the problem of killing the rats in sewers.

Everyone's nightmare – a rat escapes from the sewers through a broken drain cover, resulting inevitably in an increase in surface infestations.

The silver and leather sash originally worn by the Royal Ratcatcher to His Majesty George IV. Today it is an annual prize awarded by Rentokil to its most efficient pest control manager.

who will be notified and told to bait the area. 'LRT seems to be cutting back on its baiting', he says, 'which adds to the problem.' The rodent control operators apologize for not producing a live rat – 'They sense our presence and will remain secure in their burrows until we leave,' says David Drake. No one is really disappointed.

Stop two, ten minutes drive in a northerly direction along one of Finchley's main roads, is in the heart of Mrs Thatcher's own constituency. Backing onto a park, next to a zebra crossing and opposite a parade of shops, is a bus stop. It is a sheltered wooden stop, constructed in the late fifties, and there is seating for five people. Bus stops, we are told, are like open air restaurants for rats. The growth over the last few years in take-away food has led to an increase in the amount of food waste dumped on the streets. David looks for holes, and finds them at once, underneath the seats amidst the mud, dust and sweet wrappers. Two coke cans bent and broken, lie either side of one of the exit holes, with a torn American hamburger container, a few chips and a wrapper. There are lots of crisp bags scattered around.

People don't seem to realize that an ordinary crisp bag with a few crumbs is a treat to a hungry rat. Rats will eat anything, the plastic bag too, so dropping litter in the road keeps them alive. They make a good meal of a hamburger wrapper – they eat the paper because it's covered with grease. Look at that drink can. Rats like licking the sugar from the top of the tin. They do the same with beer bottles.

The bus stop backs onto a wire fence – on the other side of which is a small park. There is a large pigeon population here and some ducks. Mark Jervis stresses the importance of keeping open areas like these free from food waste.

Several members of the local community feed the birds, leaving all kind of scraps for them. The problem is that what the birds leave behind, the rats finish. It's hard to tell children not to feed the pigeons – but the alternative is healthy rats in large numbers.

Mark moves towards a bush where there are signs of a run.

The infestation was treated several weeks ago, and it looks like the burrows are no longer 'live'. David checks the takes; only two of the five bags have been touched, a good sign. He balances on several unstable paving stones, which rock beneath his weight. From their movement, the ratcatcher estimates that the rats have burrowed nearly twelve feet towards the road at a width of two to three paving stones.

Mark shakes his head. 'We're operating a fire service at the moment,' he admits, 'We're winning in the short term. Barnet's one of the lucky boroughs which got an increase from Thames Water this year but if the rats go on rising in number like this over the next six months, we're going to need an increase from Thames every year.'

The third stop is just around the corner. We are outside a private house in Huntingdon Road. A few weeks ago Barnet's Pest Control Office had been flooded with complaints of rat sightings around this area. David tells us that rats never suddenly appear in great numbers at a location previously free from infestation, so there is always a specific reason for several reported sightings at one location. To the rodent control officers, it bore all the tell-tale signs of a rat break-out from a defective drain or possibly a manhole. Within a week they had traced it to this address, an Edwardian house on three floors with a small, overgrown front garden. Underneath the tall grass and weeds is a manhole. The occupier of the house is an elderly woman – the owners are local estate agents.

David Drake visited the property last week and discovered that the rats had indeed escaped from the garden's defective manhole cover. Below the manhole, another cover from part of the drain which should have been blocked off had disappeared, thus allowing the rats to escape from the main sewer network and run up into the private sewer. In this case it is the owners who are responsible for the sewer, not the tenant. David checks the takes.

David likes baiting sewers. It is like finding the solution to a puzzle, his way of finding out where the hidden enemy is lurking. First, he fills a see-through plastic bag with warfarin poisoned bait. He ties a knot on the top of the bag and attaches it to a piece of string which he wedges at the side of the manhole. He suspends the bait well above the level of the water; the bait is useless if it is allowed to be washed away. Baiting serves two purposes. It is ultimately used to kill rats; it is by far the most efficient and cost effective method. Baiting is also used as a way of estimating the approximate level of infestation. The amount of bait taken over a set period from the area of infestation is compared with the average amount taken by one rat, and from that the rough size of the local rat population can be determined. Baiting is then concentrated on the high-take areas and repeated until the takes remain untouched and the infestation can be presumed destroyed.

The takes in Huntington Road have all been dragged down and eaten so we now know where the rats are. Mark instructs David to arrange for a new manhole cover to replace the broken one and to organize a new sewer cover to replace the missing one. David baits the area again and he will visit the site later on in the week, and judge the progress being made.

On the way back to the van, David and Mark point out the large amount of black rubbish bags which are piled up around dustbins in the front gardens of several houses. Mark says it is a sign of the times. We are living in a more affluent society and now throw away more foodstuffs than ever before. We are also not taking enough care of our rubbish. Rats have no trouble in eating through bin liners and black bags as the smell of rotting food is very attractive to them. The fact that most of these houses have only one bin each is an indication that they either do not know, or do not care, that they are encouraging rats into their area.

Our fourth stop is a council housing estate where rats have been a historic and recurring problem. This is a modern development built on the site of an old one which was totally demolished. The borough's aim was to rid the community of poorly constructed housing and replace it with the latest in building technology. The old estate had an acute rat problem which was built into the very foundations of the site.

Unfortunately the rats were not dealt with at source when the houses were knocked down and so the problem remains today. Mark says that the rats should have been baited to extinction and the old sewers sealed up long before the new houses were built. The problem has been made worse by modern methods of construction; the use of plastic sealers and piping has made it easier for the rats to gnaw their way back into the houses through air vents and drain shafts.

Manholes are lifted and the test baits checked: no takes so far. This is a regular stop for David Drake. He visits at least two homes each week which have reported rat sightings and the area is currently under the department's intensive treatment programme. Two more covers lifted, four lifts in all, and it is back to the van.

Our last stop takes us down the North Circular Road past Brent Cross Shopping Centre towards Friern Barnet. The site we are to inspect posed a major headache for the department some months ago, with reports of dozens of rats in the area. Today the colony is no more; it has been baited to extinction. We are here today because it demonstrates a common problem faced by local authority pest control departments. The area includes a pitta bread factory at the back of a large supermarket. There are several sealed skips containing leftover dough from the factory, due for disposal. The building itself faces a twenty foot high cement wall, constructed to firmly hold back the land behind it. The honeycomb pattern of the wall, which traps mud in all of its numerous cavities, provided perfect harbourage for rats. When the pest control team first arrived at the wall, which runs for nearly one hundred yards, they were faced with a colony of over one hundred and fifty rats. 'The slats provided ideal "rat-flats", ample space on each level for rats to live in, with plenty of food and water from the pitta factory,' said Mark. 'Old dough had been carelessly thrown on the ground outside the back doors – some of it had stuck to the wall itself – and dripping taps provided the daily supply of fresh water.'

The first action taken was to force the factory to clean up its act by tightening up its methods of waste disposal and rubbish storage. The rodent control operators then began a series of intensive baiting programmes and within a few months the entire colony was wiped out. According to Mark, the main culprits of the piece are the architects and planners who designed that particular structure of land-supporting wall. He suggests that had the builders only talked to his department before they started, he could have told them that their designs were literally the opposite of rat-proof. Mark suggests consultation is needed at the planning stages of building developments if similar problems are to be avoided in the future.

It is nearly the end of the afternoon and David Drake has several more calls to make. He apologizes once more for not having produced a single rat in our afternoon's travels. 'I suppose it means we're doing a good job!' he smiles, as he gets into his yellow council van. It is work that very few people could do.

Back at his office in what used to be the old Ever-Ready building near Totteridge Lane, Mark Jervis, Barnet's Pest Control Supervisor, sits at his desk. On his walls are posters bursting with information about rats, mice, fleas, bees, wasps, squirrels and other nuisances which are the responsibility of the pest division. Magnified pictures of cockroaches and dire warnings on the perils of the common household fly are pinned about the place. A small pest control library sits on a shelf and spills over onto the Pest Control Supervisor's desk. Next to the books, a bizarre collection of souvenir paperweights rests. Every one tells a tale with the moral – beware the rat! A six-inch piece of lead gas-piping, gnawed in half; power cables fused together after rats chewed through part of the grid causing a fire; and a green Wellington boot scarred with a dozen deep rat-bites, its owner a rodent control officer working below ground.

Facing the desk hangs a large map of the borough, covered with a pattern of different coloured pins grouped in several areas. It looks like a Second World War map indicating the positions of aircraft bombers and their targets. Mark Jervis stands next to the aerial view of his borough, his fine set of whiskers making him appear every bit like an RAF squadron leader. He shows off his baiting programmes with pride: red pins and yellow pins charting his plan's progress, as well as new sightings of infestations and treatments. His strategy is simple but effective: regular baiting of the key trouble spots, special attention to rat sighting complaints and the annual baiting of target areas covering the whole borough.

Barnet is the third largest borough in London, stretching over 22,124 acres. It shares its borders with the London boroughs of Harrow, Brent, Camden, Haringey and Enfield, and with Hertsmere District Council. The borough runs through Edgware, Burnt Oak, Colindale, Hendon and Golders Green on its east side; Barnet, Brunswick Park, Finchley and Hampstead Garden Suburb on its west; with Mill Hill, Totteridge and Arkley at its centre. Barnet includes over a dozen open spaces, including Hadley Common and Woods, Scratch Wood, the Hampstead Heath Extension, Edgwarebury Park and Brook Farm. There are several golf courses, playing fields and recreational centres, including Barnet's Copthall Sports Centre. The Brent Reservoir (also known as the Welsh Harp) sits to the west of the mouth of the M1 motorway and the borough includes sections of several busy roads, the Watford Way, the North Circular Road and the Great North Way. The London Borough of Barnet is a mixture of town and country development and is suffering from a huge increase in rat infestations.

The figures collated by Mark Jervis' department speak for themselves. In 1987–8 there were 2,620 reports of rat sightings in the borough – an increase of just 1.3% on the previous year. But the 1988–9 figure rose to 4,021 complaints, an increase of 53.4%. If the 1989–90 projection follows the currend trend, and Barnet's Pest Control Division sees no reason to doubt it, the number could be as high as 6,168 by the end of the year.

'There is definitely an increase in rats this year,' says Jervis, and he claims there are several reasons for this. The mild winters that Britain has been enjoying over the past few years have increased the number of rats on the surface. Since Barnet is semi-rural in areas, with lots of open spaces and grassy land, the borough provides excellent harbourage for rats. It is surface infestations which are the direct responsibility of the borough, and Jervis notes that while other boroughs have a problem with their sewers, he has a major problem above ground. Barnet's farmland, parkland, waterways and railways all add to his difficulties. There are more take-away restaurants and fast food chains in the borough than ever before, and Jervis sees this as a serious contributing factor. 'People don't have any pride any more. I've seen it. They come out of the pub, go and buy a burger and chips which they eat as they walk down the road. The bus comes, they lob it over a fence and just run for the bus. Half a burger, a dozen chips and some greasy paper, they get blown under a bush, a rat comes along, and the grub's there. Now the rat knows where to eat, it'll come there regularly.' The problems are made worse by the growing amount of rubbish being left in the streets. With an increase in the amount of packaging that has to be thrown away, households are using more black bags, which are often left in the open for a week until they are collected by the dustmen. Rats can and do chew into these bags quite easily.

There are more demolition sites in Barnet than ever before with the expansion of commercial, industrial and private developments. Jervis says that the builders are not looking after their drains as they should. When a site is made vacant, drains must be sealed properly, and very often this is not done. The newer sites are being constructed with plastic pipes and drains, which are easy for rats to gnaw through. The result is more rats escaping to the surface where food is ready and waiting.

Another factor is the attitude of British Rail and London Regional Transport. Both bodies, says Jervis, are responsible for controlling their rat infestations but they fail to do this. Although Barnet does pass on information of reported sightings, the general feeling is that the transport authorities do not take the problem seriously enough and are actually cutting back on their baiting programmes. In practice, it means that however well Barnet's ratcatchers do, rodents will escape onto the relatively safe areas of BR and LRT land, where they will continue to breed.

But Barnet is exceptional in one respect: it has no gripe about its water authority, Thames Water. In fact, Barnet is regarded as 'the mark by which other boroughs are judged', according to Mark Jervis. Barnet receives an annual budget which is based on treating half the Thames Water manholes in the borough at any one time. Its regular contract with Thames Water was worth £10,000–£12,000 per annum – but this year it has risen to £20,000. 'We are the best value in London,' boasts Jervis, 'Each lift costs us £1.40 and we budget accordingly.' (A lift is one manhole treatment. The cover is lifted and bait is lowered on a string. Most manholes have to be lifted at least twice before they are considered dead.) Barnet's Chief Environmental Health Officer approached Thames Water for more money earlier this year following the serious increase in rats. The water authority granted the borough an extra £8,000 to cover the increase in baiting. The money will allow Barnet to bait a larger area and keep in control of the rat problem.

As things stand, the borough is still ahead of the game – but if the rat population continues to grow in size by the same percentage as it has this year, Jervis says he will again need even more cash to cover it. It is unfortunate that all water authorities do not share the same reasonable and practical relationship with their own local authorities as Thames Water enjoys with Barnet. The majority of EHOs say that Barnet's case is the exception and not the rule and point to other London boroughs who are experiencing severe cuts in real terms in their financial allocations from Thames Water. With the money from Thames Water, Barnet contracts out to a private company called Robug. Barnet maintains a monitoring role in Robug's progress, compiling the statistics. The company baits to extinction, and reports back to Jervis who points them to another area.

Jervis 'would not mind' if Thames Water cut out his role in the operation altogether. 'Providing the new private company would continue to monitor the situation and to pass back the information here, I would see no problem with that. My only concern would be that, at the moment, I tell the company to bait a specific area – if I had to *ask* them to bait, things might be different.'

Barnet's Pest Control Division does not deal with rat infestations in restaurants. Responsibility for this is contracted out to private companies. The borough usually sends its health inspector with a team to investigate the complaint and to make a survey. Jervis then insists that a private firm be brought in to effect a treatment.

'It all goes back to an incident in Westminster, when a health inspector was making a survey and found rats present. When he asked the owners what they had done about it, they replied that they had done something – they had called the council's pest controllers in to treat it. The inspector replied that the treatment was not good enough because the restaurant still had rats. It ended up with the council trying to sue itself for not carrying out treatment.' Jervis does not want to see this case repeated in his borough.

Mark Jervis is a man who knows his subject well. Although today a supervisor in charge of the Pest Control Department working mainly from behind a desk, he made it up through the ranks from the job of ratcatcher. When a young boy, Mark used to go on rat hunts – he says he really wanted to be a big game hunter, but had to make do with small rodents instead. Before working for the council, Mark Jervis was employed on pig farms 'just for the ratting'. He'd catch them by hand or by shooting them. He has a practical and working knowledge of rats, and is particularly strong on rat bites and how to avoid them.

You often hear about people being bitten by rats, but you only get bitten if you try and pick one up. If you want to catch a rat alive, by hand, the very best way to do it is to pick it up by its tail, but hold it clear of your body. If you don't, they urinate on you straight away as a defence mechanism. You have to shake it otherwise a small rat will climb up its own tail and bite you. An adult can't do that, so it will spin itself round and round until it splits the skin of its tail off, and you get left holding a sac while the rat escapes with a sore tail.

Like all ratcatchers past and present, Mark Jervis has enough gory stories to send even the strongest-stomached listener to bed with nightmares. His most horrible tale concerns an incident which happened five years ago when he worked as one of the borough's rodent control operators. A mother had brought her three-month-old baby into the local hospital's accident unit, its bottom lip bleeding profusely. The doctors diagnosed the wound as a rat bite, and Jervis was called in to investigate. Half a mile away from the woman's house, council workers had been digging up the sewers. Jervis traced the rogue rat's path along the road to the back of the house. It had gnawed its way into the building through a plastic tumble drier vent in the wall of the utility room, and headed through the kitchen, and up the stairs.

'It was a nice pink pile carpet, I always remember that,' says Jervis. 'The door to the baby's nursery room was open, and the baby lay crying in its cot. Around its mouth were traces of milk – the baby had just been fed and had regurgitated some of its feed. The rat had climbed into the cot and started licking the milk off the baby's face. The baby must have tried to brush it off or just moved its head in the wrong direction, and the rat bit it. The story has a happy ending, smiled Jervis, 'the rat had obviously been a regular visitor to the room, so it was easy to catch and kill and the baby survived with no harmful side effects.'

There have been over 3,000 rat sightings in the London Borough of Barnet so far this year. That's in line with forecasts based on last year's increase of 53%. Barnet can be proud of its Pest Control Division which has kept on top of its rodent increase with sound administration and professional expertise. But Jervis warns of the dangers of a rat population increasing in size. More money is needed to fund more baiting programmes not merely for running a fire fighting service. The funds should come from the water authorities as well as central government. But Jervis is also critical of the public's role in facing the challenge. 'We must change some of our own bad habits, especially when it comes to litter in the streets and the way we dispose of our rubbish,' he warns, 'the very existence of rats is a threat to our good health!'

10

Manchester's Rats

In 1987–8, there were 2,725 rat sightings reported in the City of Manchester, an increase of nearly 2.5% on the previous year. Just twelve months later, that number had risen dramatically to more than 3,834 complaints, a rise of more than 40%. Manchester City, with its population of around 450,000, is currently experiencing the worst increase in rat infestations in its history and if the present growth pattern continues with no extra financial help from the water authorities or from central government, the future looks grim.

All of this comes as no surprise to Rob Mylchreest, one of the City Council's ratcatchers. Accompanied by Harry Beasley, the Council's Principal Environmental Health Officer, Rob's tour of Manchester's rat hot-spots is enough to convince anyone that there is an acute problem here. There are too many rats and too few catchers. There are simply not enough resources to go around, and it is the operators on the ground who are expected to cope with an increasing demand for their services.

This morning will be spent following up several reported rat sightings and checking last week's takes. First stop is in Bow Lane, a small back street in the City centre. Someone found a dead rat on the road a few days previously and the rodent control officers have been called in to discover whether it is a one-off event or the first sign of an infestation.

There are several holes nearby, and by the look of the exits, they are still in use. The whole area is live. Rob discovers fresh rat droppings on the ground, spots several distinctive rat smears on a white wall close by; the infestation is confirmed. As this particular case is situated near a pizza restaurant, Harry Beasley speaks to the manager to discuss the problem. The pizzeria is completely up to public health requirements and is not responsible for the rats. However, Harry warns the manager to be extra careful with the restaurant's rubbish, particularly its discarded pizza boxes. If Harry is about to bait the immediate area, he must ensure that other, perhaps more appetizing food does not compete for the rat's attentions. 'Take-away food shops can often be the lifeline of a newly breeding colony,' says Harry.

Rob discovers a drain near the rat holes which looks defective. Drainage experts will be brought in later in the week to examine for breaks or possibly out-of-sight cracks which could be allowing rats to escape. The process of checking drains for breaks and openings is simple; a dye is poured into the sewer then flushed out immediately with water. If the dye appears in the drain, then there is some kind of leakage, and the drain must be removed and replaced. For the meantime, however, all Rob can do is to bait the area thoroughly. Bow Lane will have to be visited again, next week.

The second stop is on a council estate just around the corner. An empty house, recently destroyed by fire, is due for demolition. Neighbours have complained about large numbers of rats running around the garden. The small yard is overgrown with weeds, and has become an unofficial dumping ground for broken boxes, rubbish bags, old cans and carpet. There appears to be nothing wrong with any of the drains here so Rob thinks the problem may lie in the house itself. In many cases the toilets in abandoned houses are not sealed properly and the rats use the openings as escape tunnels to climb to the surface.

The property represents an interesting problem for the council. The house is boarded up and Rob was unable to gain access to it on his last visit. As it is a council house, part of an estate, he cannot break into it without the permission of the Housing Department. Rob works for the Environmental

Health Department, and even though both departments come under the control of the City Council, Rob cannot simply smash his way inside. The building was sealed by the Housing Department after children were seen using it as a playground. Had the house been privately owned, the Environmental Health Department could have served a notice on its owners. If that had failed, the EHOs have powers of entry backed up by the courts to ensure that infestations are treated and the council paid back in full for its services. It is a fact of life, however, that councils usually do not take action against themselves, so the house remains sealed. The matter is now in the hands of Harry Beasley, who is trying to hack his way through the red tape and paperwork. But for now, in case the house is not the primary source of the problem, further urgent searches must be made for infestations around the site.

A huge board is discovered below a pile of rotting rubbish. When it is lifted, a rats' nest is found below. Several rats are killed and the area is baited. Meanwhile more nests are found in several locations in the neighbouring garden. Since there are no obvious signs of breaks in any of the other drains, the answer must be somewhere else. After more than half an hour's searching, the source is discovered opposite the neighbour's garden. The offending area used to be a rollerskate park, and the drains have been left open; gaping holes run up to the surface, giving the rats an easy climb. All of the rat infestations that have been discovered here this morning can be traced to these two open drains. Rob and Harry are frustrated with the slap-dash way of building that is common today; open drains are unfortunately no rarity and they will have to be sealed by the Council. They bait the remaining holes and nests around the drains and Rob continues his day with a trip to the other side of town.

Harry Beasley returns to his office in the splendour of Manchester Town Hall. Manchester was created as a borough in 1838, and became a City in 1853. With its population of nearly half a million, the City elects five Members of Parliament and is the Regional Centre of Government for the North of

England, with offices representing most government departments. It is a proud City, with a definite anti-establishment past which survives in its widespread opposition to enforced local government cuts in jobs and services.

There are 33 wards in the City, with Blackley and Crumpsall on its northern edge and Woodhouse Park at its southerly tip. Manchester shares its borders with the Metropolitan Districts of Trafford, Salford, Bury, Rochdale, Oldham, Tameside and Stockport. Although Manchester is a mainly urban environment, it has its fair share of parks and open land. There are 12 principal parks in the City, including Heaton, the biggest at 245 hectares, and Wythenshawe at 112 hectares. Out of a total of 28,750 acres that make up the City of Manchester, 7,312 acres are woodlands, reservoirs, farms, tree banks, open green spaces and parks.

Harry Beasley is the Principal Pest Control Inspector employed by Manchester City Council. His department deals with all types of pests from squirrels to wasps, from fleas to mice, but he gives 'top priority' to any complaints concerning rats. The standard procedure following a telephoned or written complaint is for one of the department's operators to visit the site to determine whether the complaint is genuine. If an infestation is discovered, it is baited immediately; failure to act quickly results in more rats in a short space of time and twice the problem, according to Beasley. If the operators are unsure, they test bait, in a similar manner to the Barnet ratcatchers. In the case of open sites, the pest controllers carry a plastic cylindrical container, not dissimilar to a section of pipe, into which they place the poison. The aim of the pipe is to make it easy for rats, which are attracted to small, dark areas, to find, but difficult for children and domestic animals.

When an infestation is confirmed, a report is sent to Beasley requesting a drainage inspection. He sends out a drainage team who search for defects and possible breaks in the local drain and sewer system. If there is an obvious defect, the teams repair the local drain leading from the sewer. If the drain is in a public area, it is the responsibility of the Highways Department. On private land, it is the owner who must arrange to have it repaired and foot the bill. 'What we aim to do,' says Harry Beasley, 'is to find the source of the infestation; we are not satisfied with simply baiting the area. Once you find the cause of the problem, you can stop it. In a built-up area like Manchester, I would say around 90% of rat infestations are the result of defective drainage.'

Apart from the general reasons for the increase in the size of the rat population – the recent mild winters, more rubbish and food on the streets – Beasley says that Manchester has a particular problem with its drains, which are crumbling at an alarming rate. Coupled with that, there are many branch drains leading to the sewers which have been termed 'dead'. At one time or other, they served a home, an office building or factory which has subsequently been demolished. The drains are not sealed off properly at the connection with the main sewer and the result is that rats have more free movement in the sewers to breed and to run up to the surface. 'That's why we have more surface infestations,' says Harry Beasley, 'We are giving them better conditions for breeding.' Another problem is the material used for new drains – plastic. As rats have to constantly gnaw with their teeth to stop them from growing and plastic drains are easy to damage, they pose a serious threat.

Changes in house-building practices have also contributed to the increase in rats. Beasley points out the advantages of the old methods of construction, when houses were built with intercepting traps. These were fitted at the edge of the property, and worked on the same principle as water traps found under any sink. They were equipped with a special rodding arm which was used when the intercepting traps were blocked. The traps often caused more blockages than they were there to clear, so they were eventually removed altogether from drains and pipes. 'I believe that this has made a rat's journey to the surface much easier, since there is now no water trap to negotiate.'

During the 1950s and 1960s, a huge slum clearance operation in Manchester demolished many parts of the City. It left behind large amounts of 'dead lengths' of sewers which were not blocked off. The new estates which were built on the old sites consequently still suffer from rat infestations as nests of rats live and breed in the 'dead' sewers, and come to the surface to find food.

One of Harry's worst work experiences occurred when he was called to a building in the City centre. Rats had been reported in the adjoining property and it was thought they had escaped from the building which Harry was to visit, an old property, sealed up and due for demolition. He gained entry to the premises and discovered that the building had been closed up with foodstuffs still inside. Rats had burrowed their way into the building, made nests and had bred there. 'The place was literally crawling with rats,' says Harry, who emphasized that what made it worse was to think that these rats had never seen the inside of a sewer; they were all the healthy descendants of previous generations of below surface rats. 'It was a particularly horrifying incident for me,' Harry admitted, 'because I hate rats – I've no stomach for them!'

Harry Beasley has strong views about North West Water's contribution to the problem. Manchester's City Engineers receive 50% of their money from the authority for sewer baiting: 25% comes from the Engineering Department itself, and the other 25% from the Health Department. Since Beasley believes that finding the source of the problem is most important, it is obvious to him and his department that the source of the substantial increase in surface infestations is the inadequate condition of the sewers. 'The engineers need far more money for the repair and maintenance of the sewers and for baiting. The Council will be asking North West Water for more money soon.' Harry Beasley says its 1988–9 contribution of £40,000 is 'absolutely useless' when compared with the increase in infestations with which he has to cope. It is easy to understand his frustration; with no increase in funding, it is becoming harder for his pest control division to live up to the slogan of his department – 'Manchester City Council – Protecting and Caring for You and Your Environment'.

As an Assistant City Engineer, Ian Birtenshaw is one of the

Council's officials responsible for sewer maintenance. According to the City Council's handbook, the Engineering and Surveyor's Department has, amongst other things, the responsibility under an agreement with the North West Water Authority 'for the design, construction and maintenance of both foul and surface water sewers'. There are approximately 1,500 miles of public sewers in Manchester which vary in diameter between 150 millimetres to over 4.5 metres. Most of them were built before 1870, and some sewers constructed in the early 1850s are still in use.

Birtenshaw's department has to negotiate directly with North West Water for an allocation of money to cover rodent control; most of the money is spent on baiting the City's manholes. The current funding agreement allows him to allocate two two-man gangs to regularly bait Manchester's manholes. Birtenshaw estimates that with over 80,000 manholes in the City, it would take more than seven years for the crews to bait every one. Yet despite this impossible situation, North West Water's budget allocation has stayed at the same level for the past seven years. Birtenshaw explains that many years ago the water authority agreed to fund 50% of Manchester's rodent control budget for the sewers and that allocation has remained roughly the same ever since. The 1989–90 figure, he has been informed, will be the same as the previous year, despite this year's problems of increasing rat infestations.

Negotiations with the authority are not easy. North West Water operates under a cloak of secrecy and is particularly cagey when it comes to discussing funding policy, let alone increasing its sewer baiting budgets.

'In 1983–4, the number of our manhole takes shot up by about 19% – the following year they showed a massive increase, around 44%,' says Birtenshaw. His department approached the water authority for more money, as it was unable to cope with the increased demand. Agreement could not be reached with the authority, so the Engineering and Surveyor's Department was forced to change its system, to see if more baits could be squeezed out of the old budget. It managed to increase its programme by 50%, baiting 12,000 manholes in the year. The most recent increase in the rat population, however, will not be as easy to deal with; the system is working to its full capacity and cannot cope with any further increases in demand. His department has again approached the authority for more funds, and Birtenshaw is waiting for a reply.

Ian Birtenshaw's job mainly keeps him tied to a desk, but his work activities occasionally take him into the sewers. He has a library in an adjacent office packed with 'video nasties' of sewer tunnels by the mile, filmed by remote control robot-camera, featuring the odd glimpse of a rat. One of his favourites was taken under the city centre in a sewer about three feet wide. The camera was seeking signs of a fall-in – a blocked sewer caused by falling bricks – when it came across a rat sitting on its haunches and feasting on a pile of rubbish which it had gathered in the middle of the flowing sewer, forming a makeshift island. The rat, seeing the bright lights of the remote TV unit for the first time in its pitch black sewer, seemed little surprised at the mechanical intruder. It just looked up at the camera, sniffed at the air, then carried on, fishing in the quick-flowing foul stream for little clumps of food, using the opportunity afforded by the lights to search for smaller morsels floating past in the filthy water.

Birtenshaw's most frightening experience involving rats happened when he was inspecting sewers for damage and a suspected fall-in. 'When you're going through the sewers, rats occasionally pop their heads out and just stare at you. Remembering how small some of the sewers are, you of course come literally face to face with them. It's an unsettling experience, to put it mildly! One time, I remember crawling through a tiny sewer. I was with some colleagues, and we thought we could see a minute pair of eyes reflecting our torches in the far distance. As we got nearer, we could see that it was a rat, lying on the ground, just staring at us without any movement. When we reached it, it turned out to be only half a rat – the back half of it was just skeleton, its friends had eaten it away.'

Manchester City Council has its fair share of people who are frustrated in their jobs. One phrase constantly echoes around the Town Hall – 'lack of resources'. Ian Birtenshaw tries hard to work within the imposed financial restrictions, but feels that proper funding is essential if he is to maintain the City's sewers at an acceptable standard. 'The allocation of money from North West Water should reflect the problems that we face,' he says.

In an office a few hundred yards down the corridor, Mike Eastwood, the Director of the Environmental Health and Consumer Protection Department, battles with the same problem. 'We are coping with limited resources. There is no doubt that we have not got enough resources in this city to be able to offer the sort of pest control service that we require,' he complains. Eastwood says the current increase in the size of the rat population, and the lack of extra funds available to deal with it, is only half the problem. 'There is a need to properly resource the service if we are to continue,' he says, 'but when we use the words "rodent control" it makes people think that we are, in fact, controlling them. In the past we've been lucky if we have. We are still using practices that have developed and evolved which work against us. We have in the past made mistakes which have encouraged a greater availability of places for rodents to live, and we haven't made sure that these practices won't lead to the same problem in the next ten or twenty years.'

Eastwood feels that the whole area of rodent control should be properly reappraised; the sewers should be adequately maintained and in many cases, rebuilt, while current construction and demolition work should be carried out with rodent control in mind. Eastwood says the only effective way to deal with rats is to get to the source of the problem: 'Unless we get to the root of it, someone else will be coming along in 100 years time to write another book on the rodent problem in Britain!' Eastwood claims that all this costs money and the scale of the problem cannot be dealt with by the granting of one-off amounts; the entire area needs reorganization and more financial support.

8 Mike Eastwood would like to see a programme of immediate

heavy-baiting of the worst affected sewers in Manchester and other trouble spots like local council housing estates.

> When these estates were built following the old slum clearances, the resources at the time were not available to deal effectively with the old drainage systems and provide proper new ones. It is a present problem that can be avoided for the future. Twenty years ago, the aim was simply to rehouse the people living in the slums; no one considered all the aspects of that activity. Therefore, the problems of the old mains sewers are with us today, affecting our new drainage systems.

Had more money been spent at the time, he argues, we would not be suffering to the same extent as we are today. Similarly, if we continue to demolish and rebuild in the way that we are, the rat problem will always be with us.

Mike Eastwood has strong views about the privatization of the water authorities, too. They are reflected by the slogan printed at the top of his headed notepaper. Bright red letters proclaim: 'Manchester City Council – Defending Jobs – Improving Services'. Eastwood's commitment to the city's environmental and consumer services is second to none, and he sees the privatization issue as yet another attack from a government which measures everything against a financial yardstick. In the case of the water authorities, Eastwood says, the twin demands of the profit motive and the provision of a good social service do not make happy bedfellows. 'Local authorities came into their own because they realized the need to deal with problems locally. If you go back to the days of Chadwick, local government was born on the issue of public health; its main activities were the provision of decent drainage, housing, roads and water supply.'

Eastwood cites a local example of Dr Duncan, the first Health Officer employed by the City of Liverpool. 'It was Duncan's commitment, courage and wisdom that inspired others to help him in his work to build over 30 miles of sewers in a City which had no sewers at all. They did all that then, and now water is to be sold off. I believe that water is owned by all of us. There are enough public health issues involved with this for me to say that we need it to be freely available to the whole public, not without paying for it, but without its provision and costing being influenced by market forces,' says Eastwood. 'A public company has to have a balance sheet for its shareholders to show them that it has produced a profit; that is not the way to go forward with the provision of a service that is so crucial to society.'

Mike Eastwood poses this question: if the key to the current rat problem lies down the sewers, where the rats are breeding, how are we going to convince people who are profit-led to take part in an activity that appears to be pouring money almost literally down the drain? 'The agency arrangements that exist between the water authorities and the local authorities end in 1992. After that, the individual PLCs will be able to make private arrangements with companies which offer more financially competitive terms.' Eastwood argues that the current job is being done by the City's engineers, who know the local area well, and have years of experience in dealing with the notorious trouble spots. He is worried that if an agency contract goes elsewhere, valuable local understanding would be lost, '. . . and I think that could have serious consequences upon public health.'

Eastwood stresses that he does not see profit as a dirty word – in fact in the past his Pest Control Division has successfully competed with some major private companies to win valuable contracts with several local hospitals. 'The feedback from the hospitals is that the rat situation has improved since we've been doing it!' says Eastwood. 'We won't be a commissioning agent for others because we're more cost effective than they are,' he says. 'My main belief is that the aims of this department are to maintain the health, safety and welfare of the population of this city. You cannot do that with discrete numbers of private companies.'

There are other measures, however, that Mike Eastwood would like to see implemented on a national level to help his department in its work. He says that as they stand, the laws governing problems with rodents on food premises need more bite. It is a point on which most Environmental Health Officers fully agree. As we heard from Mark Jervis in Barnet, closing down a restaurant where rats have been reported can certainly take days if the owners do not co-operate, or even longer if they decide not to take action until just before their case reaches court. On the other hand, local authorities have full powers to close an establishment immediately under the Health and Safety at Work Act by serving a prohibition notice. Mike Eastwood would like to see the same powers given to his Environmental Health Officers, powers which would enable them to shut down any establishment selling food immediately, once rat infestations have been discovered on the premises.

'All food premises with a rodent problem are an immediate and serious threat to public health. The presence of any rodents in the environment worries me, and this new increase is obviously something that I am very concerned about,' says Manchester's Director of Environmental Health. With no more cash on the horizon and next year's rat population figures set to rise by at least another 40%, Mike Eastwood has genuine cause for concern.

11

Water Rats

The rising levels of rat infestation in the London Borough of Barnet and the City of Manchester are not unique. The current explosion in the size of Britain's rat population is a national problem, and to many people's anger and frustration, its coming was not unannounced.

Towards the end of 1987, reports began to filter through from many local pest control operators around the UK that they were experiencing an increase in the amount of reported sightings. Complaints of surface infestations are generally the first indication that rats are on the increase; it is easier for people to see and report rats running around the streets than to notice any rise in the number of rats in the sewers. Since most local authority figures are not collated until several months after the end of any one year, if indeed they are at all, official confirmation of the increasing rat problem was slow in arriving. Individual local authorities complained on a purely local level, asking for more funds from central Government and the water authorities. When national warning bells should have been sounding, they did not. Instead there was a plethora of 'one-off' reports and surveys in the local and national press. The signs were there – but no one noticed them.

In September 1988, Peter Archer, the Assistant Chief Environmental Health Officer for Bristol, reported that a reservoir of rats in his sewers, an infestation level of nearly 50%, had been making its way to the surface and onto the streets by tunneling through the soft earth exposed after sewer fall-ins underground. In Camden Town, north London, dozens of rats were seen eating grease on the train tracks; the *Observer* reported one guard explaining that he had been issued with a large stick to keep the rats from getting onto station platforms. At Waterloo Station, following complaints that rats had been seen crawling over several people sleeping rough on the ground, a pest control unit killed a colony of sixty rats. Lambeth pest controllers said that the colony had escaped from the sewers through a broken drain. The infestation rate in Lambeth was very high; one controller explained that in most of the cases he visited, rats had escaped from 'crumbling sewers' and even made their way into houses by popping up in toilet bowls.

In a report to the London Chief Environmental Health Officers Association in the same year, the Pest Control Committees Co-ordinator attempted a survey to find out how serious the rat problem was amongst all the London Boroughs. Of the 24 boroughs who were able to give precise figures for surface infestations, just two reported no increase at all, while the rest confirmed the committee's worst fears – increases ranging between 5% and 73%, giving an average increase for the whole of London of 30%. 12 boroughs reported that they were buying more rat bait than ever before. Of the many causes given for the rise, the survey highlighted the increase in London's demolition and construction work in particular, with specific regard to the failure of local contractors to seal off open drains.

An increase in fly tipping, take-away food shops and the public's careless disposal of litter and use of rubbish bags were also blamed, as well as the mild winter of 1987–8 and the fact that many public landowners such as British Rail were cutting back on pest control. The report was met with general apathy.

In Hertsmere Council's annual magazine, David Ward, the Council's Pest Control Officer, drew his readers' attention to the growing problem of rats in the local area. He pointed out that although there had been a lot of press and television coverage of the problem of the increasing number of rats in Britain, nothing had been done to combat the situation. His frustration was, and still is, common throughout the pest control industry. The writing on the wall was plain for all to see, so why had nothing been done? In the meantime, Ward offered his readers some suggestions to help them at least address the problem temporarily. He asked them firstly to stop throwing their sweet and food wrappers on the ground and put rubbish in bins. Half-eaten packets of take-away food should not be left along river banks or down country lanes and waste food should not be thrown out for the birds as all rubbish encouraged rats. Ward noted that 20% of his sewer inspections resulting from rat infestations were on council estates or new housing developments where waste disposal units have been fitted. 'They're very nice for the rats in the drains,' he said, 'they don't even have to chew their food now, we grate it for them!'

Most of the media coverage of the rat problem centred around individual incidents of rat sightings and the chance of catching leptospirosis. The story of a 78-year-old woman who fell into the River Thames at Kingston Bridge, contracted the disease and died was widely reported. A rat which had escaped from the sewers and got into a council swimming pool at Elephant and Castle was found swimming in the shallow end; it was destroyed and the pool drained and cleaned. Fear of Weil's disease again dominated the coverage of the incident.

Some scare and panic stories have arisen which have done more harm than good, according to environmental health officers. The *Evening Standard* recently published a story under the headline 'Rat Disease Peril in the Park Water'. The incident centred around a fountain in Battersea Park which was fed from the Thames. Some local children occasionally played in the water, and this was worrying a few parents. Since parts of the Thames have been proven to be infected with leptospirosis, the report asked, why did the council not put up warning signs to keep the children away? Annexed to the end of the article were a few lines from a local resident who saw two large rats in his garden, and other incidents of rat sightings from around the whole of the London area. 'The article was no doubt well-intentioned,' said one environmental health officer, 'but it just worried the public needlessly. The Thames is a long river, and only a minute fraction could be considered to be infected. The journalist didn't get to the root of the problem – why are rats on the increase, and how can we stop them?'

From June 1988 to February 1989 over 150 stories on rats and Weil's disease appeared in the press, on television and on the radio. People were beginning to get a national perspective on the problem and began looking for someone to blame. To the Association of Metropolitan Authorities, the guilty party was easy to find.

According to the AMA's survey towards the end of 1988, the main culprits responsible for allowing an unchecked increase in the size of the national rat population were the water authorities. It pulled no punches, detailing several instances of metropolitan authorities which were experiencing particular problems with their local water authorities.

Croydon Council commented that Thames Water was reluctant to agree on any increase in its allocation. The meagre sum of £7,300 per year for sewer work forced Croydon to cut back on its baiting programme. Wakefield observed that up to 1988 it had experienced little problem in getting money from the Yorkshire Water Authority, but with the approach of privatization, the YWA asked for cuts in costs. Sunday working was suspended resulting in Wakefield's main roads being omitted from the baiting programme. Bury reported that its funds were officially insufficient to fully treat the sewers according to the MAFF guidelines. Waltham Forest, which had recorded decreases in its infestation levels for the years 1985–6 reported that Thames Water had decreased its funding considerably, slashing its 1987–8 budget to a third of its original 1985–6 allocation. Newcastle's rodent control budget from the Northumbrian Water Authority rose only in line with inflation despite its City Centre manholes showing an infestation level of 40%, and growing.

Harrow's sewer treatments had been very successful leading to an infestation level lower than the Thames Water Authority's recommended 5%. Unfortunately the authority then began reducing its budget. Harrow commented, 'Thames Water Authority see sewer baiting purely as an accountancy exercise, setting global cost limits per manhole without taking into account the practical considerations of individual authorities.' Harrow is concerned that the cuts could have an adverse effect on rodent control.

Knowsley is pressing North West Water directly for funds to tackle the increase of rats in its sewers. Two areas in the borough were not treated in 1987 because the council lacked sufficient resources. After an inspection programme in the Huyton area of Knowsley, it was revealed that, out of 19,000 properties, 20% of the houses showed signs of some rat activity and in some streets it was as high as 50%. Wigan, which reported to North West Water that 60% of its sewer system was rat-infested, was told that it would not be getting an increase in its budget.

The spotlight has undoubtedly hit Britain's water authorities over the last few months; the complex discussion of water privatization has brought the various activities of the water authorities as well as the quality of its water under public scrutiny and few people have been entirely happy with what they have seen. Since the Government announced its intentions to sell off the water industry, the water authorities have been hit by an unfortunate coincidence of events which have brought about collective criticism from millions of people. The rat problem has been swallowed up in the issue of privatization.

The European Commission is currently prosecuting the British Government for failing to clean up its supplies of drinking water. It is an old problem. Back in 1980 Britain joined its EEC parters in agreeing to a safety net of standards governing a myriad of water pollutants. All countries promised compliance by 1985, and yet, four years later, nearly two million people in the UK still run the risk of drinking water with unacceptably high concentrations of lead and aluminium and a further 1.7 million risk drinking water with excessive amounts of nitrate.

The European Commission's six legal actions over Britain's

low quality of drinking water come hot on the heels of another impending action from the Commissioners over the unacceptably low quality of water at Britain's beaches. The cases which the Commissioners are currently prosecuting are: excessive nitrates in water in Redbridge and Norwich; aluminium in water in Birmingham; aluminium and bacteria in water in Bradford; aluminium, bacteria and coliforms from sewage in water in Calderdale; as well as unfair methods of measuring lead content in general in the UK. (The EEC measures lead content by running the tap and measuring the water. The British Government argues that the water should be measured after the tap has been allowed to run for a few minutes, to allow the lead from the piping to be flushed away.) The Commission is currently known to be considering a host of other additional complaints concerning British water standards.

During the Water Bill's passage through Parliament, Ministers told Parliament that the water authorities could not reach the agreed EEC standards until 1995 – ten years too late – although pressure is now mounting to bring that date forward. The cost of bringing the standard of water up to scratch is huge, and would have been bound to put off potential investors in the new private water companies. The government solved this problem by promising to send water to a public marriage with a substantial dowry to cover the bill.

The hot summer of 1989 saw many authorities fail to cope with the increased demands of their customers, exposing inefficiencies caused by a lack of capital investment. Water consumers were told that the water shortages were due only in part to a lack of water; the pumping equipment was old and ready for replacement and they simply could not cope with the extra volume of water required. In some areas of the UK, water shortages meant a hosepipe ban and other areas had to suffer with no domestic supply at all, relying on street pumps. As the water authorities reeled under the mounting pressure of bad publicity, they were hit by a new blight – outbreaks of water deemed 'unfit to drink'. People discovered 'worms' in their drinking water; representatives of various water boards drank cups of wriggling green water on television in an attempt to prove that the larvae were completely harmless, but for most, the very thought of drinking small and moving creatures was enough to put them off. Complicated radio announcements informing customers to boil all drinking water brought yet more negative publicity for the Water Authorities.

In several more unrelated incidents, the increase in algae caused by the hot summer forced the water authorities to close down several reservoirs. In one case, a sheep died as a direct result of drinking water from a reservoir, and public opinion of the water authorities in general dropped to a new low. Criticism was also raised at the large amount of money which was used in promoting the privatization campaign in television commercials and on billboards. In an *Observer*/Harris poll taken in September at the height of the campaign nearly half of those questioned found the privatization advertisements 'annoying' and a third said that they were particularly irritating at a time when there was said to be a water shortage. The Government's £40 million campaign to sell off the 10 Water Authorities of England and Wales was described by the *Observer* as 'an unqualified advertising and marketing disaster'.

In short, the Water Authorities have come under attack from all sides in the last six months. Unfortunately the finger of blame pointing towards them for allowing the number of rats to increase in the sewers has been seen by the authorities as just another argument in a political battle against privatization.

Many local authorities, however, do not see it like that. They have been complaining about the rise in rat infestations for the last five years and now, with the increase the highest it has been for over a decade, they are looking to the water authorities for an urgent injection of cash. If privatization means anything to them, it is a further cause for concern that they may face even more cuts in funding in real terms when the authorities become private companies and that rodent controllers will not be able to do their job.

The London Borough of Tower Hamlets is a good example of a local authority suffering from increases in rat infestations as a direct result of the failure of its water authority to provide adequate funding for sewer baiting and maintenance.

Len Lloyd is the Senior Pest Control Officer for the borough which includes the Isle of Dogs, Wapping, Stepney, Bethnal Green, Globe, Bow and Poplar. Len has been in the job since 1971. At that time, rat complaints were at the highest level ever known in Tower Hamlets, and were treated on a 'one-off' basis; the sightings were reported to the council and the rats were destroyed. When Len took over the job as Senior Control Officer, he instituted a system which he called 'Black Spot' control. It was a method of prevention as well as treatment, and meant organizing the specific baiting of key areas over one year to keep rat infestations at bay. The Black Spot system was a complete success.

'In those days the GLC had hundreds of properties going to waste in the borough, a great source of harbourage for rats.' Len put pressure on the GLC to demolish the houses and this, together with his new prevention treatment, saw the number of complaints drop dramatically from 12,000 in 1971 to just 121 in 1985. 'Prevention means looking for the reason for the problem, then getting it put right. For example, if it's a broken drain or an old drain, get the owner to fix it. Rat problems should be resolved once and for all, that's the nature of them.' Len Lloyd is responsible for 12 operatives and three pest control officers. His teams have worked with Thames Water crews in the sewers.

Thames Water owns London's public sewers. Each local authority works as an agency for the water authority, carrying out work within their boroughs like disconnections, new connections and general maintenance, everything to do with the sewers including blockages. Thames Water gives an allocation of money to the boroughs according to what it estimates to be their needs. This year's allocation of money, we have been told, is going to be the same as last year, which was below the level required.

Thames Water, like many other water authorities, does not discuss the financial basis for its decision-making over its sewer contracts. Len Lloyd explains that it is all a question of lifts.

The contractman will lift a manhole cover, hang down bait, then replace the cover; that's called a lift. According to Thames Water, the process should cost £1.40 per lift, one pound for lifting and replacing including bait, and the remaining 40p for administrative costs – supervision and management. Several years ago, Tower Hamlets was doing 19,000 lifts per year for about £26,000. Since that time, of course, it's much more. The cost of materials has gone up. Labour is more expensive, and that forces up the cost of both doing the job and the administrative fees. We now do it for £1.83 per lift. So, from the start we are under-financed!

Len says that increased transport and depot charges have helped force prices up, and that the new figure of £1.83 still represents good value, as it is roughly in line with inflation.

What's happening now is that only 15,000 lifts per year are being done. For example, one baiting programme is required per year. We go through 6,500 manholes, giving them one full treatment each. We break the borough up into four sections covering what was Stepney, Bethnal Green and Poplar. We do a full programme, and then get an idea of where the infestations are; you can see a pattern of takes in different areas. We go back to the infested manholes and check again within so many days. If it is still infested, we go back and do it again until it is baited to extinction.

The information is transferred to my master sheets, which gives us a pattern of where the problem areas are; some never have a problem, others always have one. This way, it is easy to see the cause of a new rat infestation. Then I produce a Black Spot system for the manholes which were initially infested on the first programme. I make it into a shorter treatment, and visit the sites three more times as part of the same programme. In that way it is thorough. I catch the young rats before they start to breed. That's the key – getting the young rats and keeping the population down. The fact is that Thames Water has kept its budget allocation down for the last three years and I can't do all the baiting I want to.

My costs have all increased; prices and wages are up and I need Thames Water to take that into account. My concern is that I can now only do two Black Spot programmes and visit 4-monthly instead of 3-monthly. This is allowing the young rats to breed and spread, so in fact the Black Spot programme actually needs to expand. I've asked Thames Water for an increase of £4,000 to cover it, but have still not had a reply.

Len Lloyd defends his department against the pressure to contract work out to private companies. He says that the current outside tenders are cheaper at around £1.40 per lift, but he feels that they are just keeping the prices down to get the work after which time they will put them up to a more realistic figure. In common with several council employees involved in rodent control, he cites a recent Monopolies Commission decision over price fixing which found one private rodent control firm unfairly competing. 'What's important is, however, that Thames Water is using their prices as a general standard to go by.'

There are, of course, other factors in the equation which account for the increase in rats in Tower Hamlets, some of which are specific to the borough's proximity to the River Thames. The Docklands development has caused the escape of hundreds of rats from the sewers and has helped provide them with harbourage. Because the Port of London Authority was responsible for running the London docks when they existed, Tower Hamlets has no accurate records of where the main and smaller sewers are located and the developers usually do not know where they are until they hit them with their JCBs. The London Docklands Development Corporation, now responsible for the area, is finding the same problem. There is also a large amount of water in the area, which comes under the responsibility of the Port Health Authority, which in turn is run by the City of London. There are many rat infestations especially near the floating offices and barges. Tower Hamlets baits on land while the City of London Corporation baits the buildings and barges on water.

A frequent complaint, according to Len Lloyd, is that old buildings are knocked down to make way for large developments without any attention being paid to the drains. No one is bothering to seal them up. As Len says,

A good example is Westferry Circus, there's a massive hole there now. Someone should have sealed up the drains. They must have thought, 'Well, if they're taking up the old roads and pavements, there's no point in sealing up the drains!' They probably thought of the money they'd save. The trouble is, even if no one's going to use the sewer again, it has to be sealed or the rats will escape. We used to get notification if someone in the borough was demolishing a building. I'd send round a team who would seal up the drains. It's not done any more.

Len says that we are now in the same position that we were in thirty years ago and repeating our mistakes. In the sixties, drains were sealed up with a bucket of sand and cement, quickly knocked together and plastered onto the hole. Rats easily gnawed their way through the makeshift sealings. Len says drains must be properly sealed if they are to stand the test of time.

Here we are, knocking things down left, right and centre, and just not thinking ahead. Modern methods of building aren't much better, either. Above ground drainage used to be made out of cast iron, all the drains were. Today everything's done in plastic, which doesn't take any time for a rat to break through. There was a case I remember of a woman who found out that she had rats staying with her because she got her feet wet. The rats ran up a plastic gulley at the back of her house – they used to make them out of lead – got into the wastepipe and gnawed its way through the plastic trap and into the kitchen, from where it chewed through the sink pipe and got into the house. The woman was washing up in the sink, when her feet were soaked with the water. That's when she discovered she had visitors!

The borough has also been affected by cuts in other public authority spending; where rodent control operations were usually contracted to Tower Hamlets, they are no longer undertaken. London Regional Transport and British Rail used to ask Len's department to bait their properties and that is no longer the case. 'I don't think they do it any more – hence we see more rats. We used to have a contract with British Waterways as well, by the banks of the river, now that's stopped, and as far as we know no one is doing it.'

Most surprisingly of all, Thames Water itself appears to be cutting back on its own programmes. 'As I said before, we bait to extinction in our sewers, yet the rats come back. Where do they come from? The main trunk sewers which are tied in to the Local Authority's sewers. They are not being baited. We used to do it with Thames Water – that's no longer the case – so now there's a back feed.'

The weather, especially Britain's recent mild winters, have had a lot to do with the increase in rats, says Len. Cold winters do kill off many of the young at surface level, but the climate is just exaggerating a problem which already exists. He is convinced that at the root of the increase in rat infestations is a more fundamental financial problem. More funds from Thames Water would go a long way in solving his troubles by allowing him to continue his proven Black Spot programme.

The water authorities, like many public utilities, are bad communicators. Cagey and secretive at the best of times, they operate in many cases oblivious to the outside world. They have been particularly evasive about the effects of privatization on rodent control, sewer baiting and maintenance. It has been suggested that, had the authorities not been seeking a public listing on the Stock Exchange, their aloof and dismissive manner towards criticism of any kind would have continued unabated. Privatization has forced the authorities to stand in the spotlight in full view of their customers and potential share-holders. Yet for all the new publicity the message appears to be the same: 'we have been at this job for a long time, we know what we're doing so don't question us about it', a case of old wine in new bottles. Local authorities working in complete accord with their water authorities are as rare as black rats in the sewer.

At Thames Water, Tony Aberdien, an acting Agency Manager with the Authority's sewerage group, is reluctant to answer direct questions, preferring to address the general issue. Thames Water, he explains, is divided into four operating areas for the purposes of sewage and most rodent control work. Each area has an agency engineer who looks after the appropriate local authority. The Pest Control legislation of 1949 gave local authorities responsibility for rat infestations and in 1974, legislation required that all local authorities were given the first option to look after the sewers. That could all change in 1992, after the full effects of privatization are known. Aberdien feels we will just have to wait and see how these changes affect the local authorities.

How does the Water Authority answer criticism that its allowances are too small to do the job properly? Aberdien explains that the allowance for rodent control is only part of the budget for sewer maintenance, rebuilding and works. Rodent control, he says, represents only 4% of the total operational budget for local authorities. If, however, the local authorities manage to save from other parts of their allowance, they can use it as they think best – on sewer baiting, for example. How much money has been given over the last ten years to local authorities collectively for sewerage works? 'I think that would be difficult to find out. There used to be nine operating divisions, now there are three, the data would be difficult to collate.'

This year's figure is £900,000 but, with nothing to compare it to, it is impossible to see if Thames Water has increased or decreased its local authority allocation. The general view is that it has remained the same.

While Thames Water does not admit that it underfunds any local authority or that its old methods of cash allocation were random, unequal and made on an ad hoc basis, Tony Aberdien says that there is a new system soon to be in operation which should ensure an adequate control of the rats in Thames Water's sewers. The authority has 48,000 kms of foul and surface sewers under its jurisdiction. The rats mainly live in the foul and combined sewers. The new scheme will aim at keeping rodent infestations in the sewerage system at a general level of lower than five per cent and to not more than ten per cent in specific areas. Local authorities will be asked to carry out test baiting to assess rates of infestation and to bait a selection of manholes that they consider to be representative of the overall rate. 'From the number of takes, a rate of infestation will be calculated; for example, if five per cent of the baits are taken, the rate of infestation is five per cent. Once the test baiting is done, they will be fitted into set categories,' explains Aberdien.

Thames Water's annual work targets will be as follows:

Category A: 30%+ infestation	– full treatment of area twice.
Category B: 10%–20%	– full treatment once plus a sum for limited retreatment.
Category C: 5%–10%	– full treatment of a third of the area plus a test bait.
Category D: 0%–5%	– full treatment of a quarter of the area plus a test bait.

The rates will be paid on the following basis: £8.70 for category A; £4.85 for category B; £1.25 for category C; and 85p for category D. For a test bait of 0% infestation, the lump sum of £750 will be paid, or the actual costs.

Aberdien is confident that under this new system, all boroughs will be allocated enough money according to their level of infestation. When challenged over the specific case of Tower Hamlets' allowance for this year, Tony Aberdien gave a general reply with all the diplomacy of a seasoned politician:

> It may have happened under the old system, but it shouldn't under the new one. I have no information on hand on Tower Hamlets and therefore I cannot comment. But I must point out that there have been cases where local authorities have contacted Thames Water and said that they had high infestation rates and needed more money. It has then transpired that the local authority hadn't been spending its allowance over the last three years!

Is there a large increase in rats in Britain's sewers this year? 'We cannot estimate what percentage of rats live in the sewers, because we have no knowledge of the number of rats on the surface,' Aberdien answers, 'but I understand from what I regard as responsible environmental health officers that there has been a considerable increase in surface infestations and therefore one might assume that there is an increase in the sewers.'

However brief and evasive were his replies to specific enquiries, at least Tony Aberdien at Thames Water was willing to discuss rodent control. At North West Water Authority, the Chairman Dennis Grove does not grant interviews on the subject. His press officer, a former journalist himself, explained with a trace of embarrassment that he could not help with any enquiries concerning the City of Manchester, with any other local auʨority, or on the issue of rats in general. 'We do not keep any files on rodent infestations,' he said, 'we do not keep records on specific budgets for sewer works either. There is no general budget for sewer infestations. The annual money allocation is made to each district on a cash limit basis.' As a matter of policy, North West Water would not reveal its total budget allocations for this year – which are hardly a state secret – and its press officer explained that it was not prepared to discuss the matter further.

Since matters of rodent control and sewer maintenance are closely linked with public health and thought to be an issue of public concern, it is surprising that a public utility, soon to be a public limited company, should treat enquiries with such contempt. It is hard to consider that the question of water authority funding for rat control and sewer maintenance is such a private matter that it cannot be discussed and harder still to believe that a water authority does not keep records of its rodent infestations and treatments. (Further enquiries should be addressed to the Chairman, Mr Dennis Grove, at North West Water, Dawson House, Great Sankey, Warrington, WA5 3LW.)

According to the Institution of Environmental Health Officers survey, three quarters of responding local authorities said that their water authority's funding had not increased to meet the growing problem of rat infestations. Water authorities may appear to work closely together in television advertising campaigns, but they act as separate units when operating their businesses. Each has a different way of allocating funds and directing work and every one has a different policy on talking to the public. A little glasnost and discussion on funding and sewer baiting, perhaps on a national as well as local level, would be an invaluable first step in confronting the problems of the current rat increase. Without the close co-operation of the water authorities and the local authorities, rats will continue to breed, indifferent to the public or private status of their homes in the sewers.

12

Rats-a-Rising

Both the public and private sector are monitoring the current increases in rat infestations in the UK. Although no official government body exists to gather such information around Britain and publish its results, the Institution of Environmental Health Officers has attempted, by commissioning a survey earlier this year, to bring the issue of rats and public health into the national spotlight.

'It ought not be too scare-mongering to suggest that all the ingredients exist for there to be once again, some time in the future, a major outbreak of infectious disease that is essentially rat borne,' says Bob Tanner, Chief Executive of the IEHO. Tanner is a man not prone to exaggeration. As the former Chief Environmental Health Officer, Tanner has had years of practical experience dealing with problems associated with rats, particularly in the area of public health, and he takes the responsibility of his current position very seriously indeed. 'We have always had rats and it is argued that rats will outlive us all as a species. Bubonic plague which was instrumental in the Black Death of the fourteenth century in London is still present in the world. With the ease of international travel, there is still the risk that it may happen again. The rat is still, as it has been for many years, Public Enemy Number One. One has to be aware that it is existing in increasing numbers in many parts of the country.'

It was Tanner's idea to commission the Rats Survey earlier

this year. His aim was to discover whether the many local complaints coming from his members about the increase in the rat population had any national significance. For Tanner, rats represent a major public health risk.

Rats always present a serious threat to the public health. Their numbers and situation at present indicate that there is a problem. To get rid of rats completely is impractical, but it is up to public health officials to keep rats at bay, keep their numbers to a minimum and be ever on guard. There are many diseases which are rat borne, including salmonella and leptospirosis.

Tanner also points to the commercial risks that rats pose – the large amount of crops destroyed in the UK each year, goods spoilt in transit and animal foodstuffs forced to be thrown away through contamination. 'Rats will eat just about anything and gnaw on lead piping, cables and drains. The cost of replacement is vast. Rats also undermine the safety of buildings, particularly in the country.'

There are several reasons why we are currently suffering from a major increase in rats, but Tanner feels that, had other factors, like proper sewer maintenance and baiting, been adequately dealt with before now, these new circumstances would not have had quite such a dramatic effect on the size of the rat population.

The recent climatic conditions have been ideal for rats. They have been able to continue their breeding cycle throughout the year, instead of having a dormant period, so most of the rats that are born survive to maturity. Harbourage is easily found – the sewers are ideal places for rats – and because of cut-backs in maintenance and repair work in the sewers, there are many breakages and cracks in which rats build their nests.

Tanner says that the reduction in the amount of work going on below grounds means that rats are largely left alone when they previously faced an annual or half-yearly scourge. The more rats are left to their own devices, the quicker they settle and breed.

Another reason for the increase in rats can be put fairly and squarely on the public's shoulders, says Tanner.

We are still, sadly, a dirty nation in comparison with others. The eating of take-away food and discarding of left-overs in the streets, the cooking and eating of barbecues in the garden. We throw away food waste and wrappers with total irresponsibility as a nation. And local authorities, as a result of Government cut-backs, don't have the funds to collect and dispose of it adequately, so rubbish builds up and provides endless harbourage for rats.

He blames the switch from dustbins to black plastic bags as another reason for more rats on the streets.

Bags which replaced conventional dustbins are not adequate containers of waste. Dogs and cats often chew open the bags, upsetting the refuse and encouraging rats to run freely through it. We're a more affluent society today, and households often have more bags than they used to. Many things that we throw out today, other generations would have kept, like bones and other left-overs which made soups and stews. Although it is a hygienic practice to throw food away rather than keep it for too long and run the risk of bacteria building up, black bags are not a good way of doing it. In short, rats are having it pretty good at the moment!

More fundamentally, Tanner argues, it is the water authorities which have not been pulling their weight, and they have no plans to do so in the near future.

The water authorities, who are responsible for the sewers' treatment, maintenance and repair, have also not been sufficiently funded to carry out rodent control to the optimum level. They will now have to look financially attractive to those who might buy their shares, and to spend a lot of money on rodent control, a rather unglamorous, difficult-to-measure-in-terms-of-cost-effectiveness activity, will have to take a lower priority in the order of things.

When the water authorities become private companies, Tanner predicts, they will continue to operate in much the same way as they did before towards rodent control.

I suspect that they will continue to do something, but conceivably they could decide to do nothing and the local authority would then have to take on the burden of responsibility to control rodents. If the private companies do not keep their infestation rates down to acceptable levels, local authorities will have to serve notice to require them to take action to reduce their rat populations, just like any other landowner.

Bob Tanner recalls the frustrations of working with limited resources on a rat problem which could be easily controlled. 'It really is appalling,' he says, 'that we know how to kill rats, how to eradicate current and future problems, but we are just not doing it.' He remembers the days when he was a Chief Environmental Health Officer with affection. Like most people whose work brings them into contact with rats, Bob Tanner has a particularly gruesome tale to tell which happened a few years ago.

During a stint in West Sussex where, amongst other things, he was responsible for rodent control, Bob Tanner was asked to attend a case at his local mortuary. An elderly woman had been brought in, and because of the unusual condition of her body, Tanner was asked to investigate the circumstances of her death.

It transpired that she had died in her sleep. She had lived in a small flat above a lock-up garage and had led the life of a recluse – so her absence had not been noticed. One morning, weeks later, in the garage below, the owner discovered a dark, sticky substance on his car, dripping from the ceiling above. He also noticed an increase in the amount of rats in the area. After the appropriate authorities had been called in to investigate, it was revealed that, after the woman had passed away, the flat had become heavily infested with rats. It was congealed blood which had seeped onto the car below – the rats had eaten away several of the old lady's fingers.

The Institution of Environmental Health Officers is based in London, operating from a building called Chadwick House. It is named after the Victorian reformer, Edwin Chadwick (1800–1890), regarded as founder of modern public health in Britain, and a supporter and encourager of up to date methods

of plumbing and sewer building. It is largely thanks to
Chadwick's efforts that we have today's sewer systems. What
would Chadwick have made of the present rat situation? Bob
Tanner has no doubts.

> I think he'd be appalled at what he'd confront. In many respects
> he'd be proud of public health standards in this country – he'd
> probably say there should be more environmental health officers!
> But I'm sure he would shake his head and tut at what he would see
> as the appalling irresponsibility that we as a nation are guilty of,
> because of the continuing enormous presence of rats in Britain. In a
> few years time we'll be entering the 21st century and we're still
> dealing with crude public health problems like rats that confronted
> our Victorian ancestors, which caused them to create the very stuff
> of local government. It is very sad that we can put people on the
> moon, we can whizz around the world in 36 hours and yet we can't
> kill rats in the drains!

A few rooms down the corridor is the office of Graham Jukes,
the Undersecretary of the Institution of Environmental Health
Officers. He views the current problem of rat increases in the
light of what he calls 'the Government's general thrust to restrict
the functions and resources of local authorities'. Jukes explains
that central government's long term plan is to depoliticize local
authorities and break them down into smaller units which will
provide only the most basic of services. Councils will be
discouraged from supplying the services themselves, and will
instead act as agencies offering contracts to private firms. In
effect, Jukes claims, the role of local government will change
into one of an agency whose job it will be to assess contracts and
enforce compliance.

> Local government is going through a massive change at the
> moment. It seems that it is shrinking back to what were its basic
> functions – control of public health, libraries and so on. I think the
> Government's view is, once the new functions have been agreed, if
> you have a local problem, you tell the Government what your plan
> of action is for dealing with it, and it will then give you the money to
> do the job.

Until that point is reached, Jukes explains, there are two bodies which could solve the current rat crisis; first, local authorities which are having to cope with an increasing amount of surface infestations and a decreasing amount of cash to treat them with and secondly, water authorities which are not spending enough on sewer baiting and proper maintenance.

We are still reaping the benefits of the investment that the Victorians put into our sanitary infrastructure. What we have found over the last ten years is that the water authorities have not been investing large amounts of money to build new schemes or to repair and make good the existing ones. They have been fire fighting, coping with problems as and when they arrive, like dealing with collapses in the middle of the road where a bus has fallen through, as opposed to instituting big programmes of investment. Most sewers were not made to bear the weight of our current traffic. While everything seems to be going well on the surface, unseen drains are not worried about.

It is those drains which are in need of repair that have allowed rats to escape from the sewers. Cuts in maintenance programmes, according to Jukes, are just catching up with us and rat baiting programmes are a part of that problem.

Over the years, the budgets have lessened. In the seventies and early eighties, regular sewer baiting kept the rats at bay, as you have seen in the case of Tower Hamlets. If you have a reduction in real terms in allowances for sewer baiting, you cannot do as many lifts as you want to, or put people in as you should be doing. With a reduction in maintenance of sewers and drains more collapses occur in the older Victorian sewers; that provides more and more ways for rats to come to the surface.

Graham Jukes explains that it is a vicious circle. Rats arrive on the surface through broken drains. Local authorities who are already unable to cope with the increase in surface infestations – their hands being financially tied behind their backs – find they have yet another wave of rodents to deal with.

> Now that there's going to be a profit element in the Water
> Authority's remit, where does that fit in with public health controls?
> That's what private companies are there to do, after all. Not just
> provide a service, but a profitable service. The whole business of
> sewers and water is basically a service which costs a tremendous
> amount of money. It's a huge investment to make.

Graham Jukes wants to see the Government pass legislation
which will force the new private companies to provide adequate
baiting treatments in the sewers, and guarantee a full future
programme of investment.

For Peter Bateman, the political aspects of rodent control
have no place in his professional war against the rat. He has
been carefully watching the current increase in rodent activity
in Britain and the controversy over water privatization with an
experienced eye. As the Director of Public Relations at
Rentokil, Britain's biggest pest control business, he says his
aims are simple: to provide a fast, efficient and good value
service for his customers. 'I leave the politics to the politicians,'
he says with a wry smile, 'it's safer that way!'

It is an understandable position to adopt. Rentokil offers its
pest control services to any organization, private or public,
willing to pay its price. The company is currently working with
several local authorities which have contracted its rodent
control services. Rentokil also sells rodenticides to local
authorities which do their own baiting. If the newly-created
water companies decide to offer their sewer baiting obligations
up for tender in 1992, Rentokil will be there, jostling for
position with the rest of them. Whatever the political climate,
Rentokil Plc., with worldwide sales of £200 million and a staff of
12,000, is bound to prosper. Like the rat, the company has a
history of survival.

Rentokil started life as the British Ratin Company in 1927,
the result of an amalgamation of a Danish research company
and a finance house from Copenhagen. Its eccentric chairman
for almost thirty years was Karl Gustav Anker-Petersen, author
of the paranoic book, *The Menace of the Death Rat*, to which we
have previously referred. Denmark was the first country to pass

anti-rat legislation, back in 1908, and has always taken rodent control seriously. During the Second World War, when Britain was facing blockade from German 'U' boats, the Danish resistance passed much needed rat-killing compounds through neutral Spain to British officials. Around that time, the British Ratin Company bought a Spitfire and gave it to the Free Danish Air Force. In 1957, the company bought Rentokil and adopted the name for its whole operation. Rentokil now covers everything from sanitary towel disposal to rat extermination, the deep cleaning of toilets to flea elimination, in premises from pig sties to palaces, in places as far apart as Birmingham and Barbados.

Felcourt, Rentokil's head office, is situated in idyllic rural surroundings in East Surrey. The property used to belong to Sir Robert McAlpine and was used as an army intelligence headquarters during World War II. Rentokil bought it in 1949. It is the most unlikely choice of building that can possibly be imagined for a company which makes most of its money killing pests. Set in acres of fields beautifully planted with trees, shrubs and flowers of every description, it is hard to believe that behind its ancient wooden doors, next to shields boasting coats of arms, underneath Felcourt's original beams, a small army of white-coated scientists huddle over their collections of wasps, moths, mice and cockroaches. In an office which could rival rooms in some of the grandest stately homes in England, Rentokil houses its administration block and scientific research laboratories. Felcourt has a nationwide reputation amongst rattophiles. 'Don't forget to ask to see Felcourt's colony of black rats,' said one of the keepers of small mammals at London Zoo, 'it's the only one left!' 'Make sure you look at the collection of cockroaches – best I've ever seen!' said a Northern environmental health officer.

If the building seems an unlikely choice for Rentokil, so is Peter Bateman. Were he not Rentokil's Director of Public Relations, one could be forgiven, on meeting him in the street, for thinking that he was an amiable professor with an interest in *Rattus rattus* and *norvegicus* that could almost be described as eccentric. Peter Bateman has one of the best collections of rat related books in the country and his easy handling of dates and diseases associated with rats marks him as an expert beyond the call of his work. If Bateman was a contestant in Mastermind, it is not hard to guess the subject on which he would be answering questions.

Bateman says there has been a significant increase in rats this year.

When I compare this year with an average year, it is easy to see that there's been a rise in the number of rats. We have noted an increase in calls – on average, up around 20% on last year. Our sales of rat poison to farmers have doubled this year on last year – that should tell us something. It also must be said that there's an increase in victims of Weil's disease this year too – about 18 people died from catching leptospirosis.

There are many reasons, in my opinion, for the increase. The mild winters over the last two years have had a lot to do with it. Rats have been breeding throughout the year; normally a cold spell drives them indoors, but this warm winter has meant more rats actively breeding and with less infant mortality, more of them are surviving. There's more food and harbourage about, too. In our towns and cities, there's more food on the streets from fast food restaurants, and now that councils have asked us to leave rubbish out in rubbish bags ready for collection, it's asking for trouble. Refuse is now more readily available for rats – bags are easily chewed. In the country, of course, farmers are cutting down the hedgerows, so there are less predators to keep the number of rats down – and the barn owl is becoming a rare breed these days.

Bateman is critical of today's methods of building and property development. He says no one is thinking of the future by taking into account rat-proofing.

Builders are not making sure that sites are completely rat-free before construction. In many cases, the designs couldn't be better for rats; they build developments which incorporate perfect homes for rats. New buildings often find rats throughout them before they have even opened because workmen bring food onto the sites, and drop it on the floor.

Bateman says he has no political axe to grind in the case of local authorities and the Water Boards. But he says there is an urgent problem below street level, and some decisions will have to be made soon.

> There is ineffective rat control in the sewers. Before 1974, local authorities had their own responsibility for rats in the sewers, and a lot of them did a good job keeping rats down. Since the water authorities have been in charge of the money, many local authorities have claimed it has not been enough and they need more for baiting and sewer maintenance. Neither is being done effectively, hence the population in the sewers is building up. There is a total lack of co-ordination in London as well. While one borough clears its sewers of rats, its neighbour might be infested. Unless it clears out its sewers at the same time, the clean sewers become re-infested.

Peter Bateman and Bob Tanner are men with different responsibilities. Tanner's primary concern is for public health in general, and his Institution's members in particular. Bateman's loyalties lie with Rentokil in general, and his company's profits in particular. But both men share a common interest – they are both concerned with the current dramatic increase in the size of Britain's rat population.

Like the majority of people involved in rodent control in the UK, they want to see the problem tackled now, before it gets any worse. We have read of the effects of the Black Death in Europe in 1348 and seen how dangerous a large reservoir of rats can be. With more than half of the country's rats infected with leptospirosis, we have reason enough to worry if their numbers continue to rise. The technical battle against rodents has been won. We know how to kill them and are capable of exterminating them in large numbers. Yet for many reasons, as we have seen, the job is not being done. There is no national body monitoring the situation, no educational campaigning about the dangers of Weil's disease or street litter, no more money on offer to match the increase in the number of rats. In the next chapter we examine some possible solutions to the problem of our increasing rat population, looking at short term answers with long term implications.

13

Towards a Solution

In attempting to find a solution to the rat problem, it is necessary to divide rats into two groups, those living below the ground, and those living on the surface. The distinction is important because it determines who, in turn, is legally responsible for their extermination. Rats in the sewers, as we have seen, come under the control of the water companies, while those on the surface are dealt with by the local authorities. No one knows the exact proportion of rats in either location, but most experts agree that the majority of rats live underground.

The water authorities can be severely criticized for underfunding sewer-baiting and operating a fire-fighting approach towards sewer maintenance. They can also be accused of compounding their problems by working under a cloak of unnecessary secrecy with little and, in some cases, no consultation with local authorities.

A strong commitment must be made by the water companies to re-invest in our Victorian sewer system; a major injection of capital investment is long overdue. The companies will have to embark on an immediate programme of maintenance and, where necessary, replacement of old and rotten sewers, with the aim of their activity being prevention rather than cure. Manchester's Director of Environmental Services, Mike Eastwood, describes the amount of money given this year to his engineers to carry out vital sewer maintenance as 'peanuts' and his Pest Control Supervisor, Harry Beasley, called it 'an absolute pittance'. Their estimation of an adequate sum for proper sewer repair and maintenance is around £450,000; at present they receive just £40,000 from North West Water. 'If we are going to get this rat problem under control,' says Beasley, 'we have to get to the heart of the matter, not just continue to bait.'

Water companies should immediately increase their allowances to local authorities, to prevent a further escalation in the number of rats on the scale we are now experiencing. As we have seen, some local authorities are dangerously below levels of funding at which adequate baiting can be provided. In fact many were operating below par even before the current crisis.

Water companies should also be encouraged to work in closer co-operation with their local authorities and to standardize their operations throughout the country.

The Institution of Environmental Health Officers suggested that if water companies were to specify the level of funding available for pest control on their land at the beginning of each year, local authorities and private companies could make clearer plans for their sewer and land baiting programmes. There is no reason why the water companies should not be more open with their figures, and publish exactly how much they spend on rodent control activities in their annual accounts.

Furthermore, if water companies worked more closely with their local authorities they could produce a standard formula for the allocation of money for sewer baiting and other pest control activities. The formula could follow the lines of the new Thames Water arrangement, but should also take into account local circumstances, the need to keep in line with inflation and predicted increases in the rat population. In this way, the ad hoc levels of allowances given to some local authorities can be monitored nationally, and a fairer system of funding developed.

The effect that privatization will have on the old water authorities has been the subject of fierce debate both inside and outside the water and sewage industry. It is not always easy to separate genuine concern from political rhetoric, and with so many forces at work on a range of practical and moral issues from the ethics of public health to the quality of Britain's water, the problem of rats has to some extent been swamped.

There is no doubt that a major programme of capital investment in sewers will not feature high on the priority lists of the new water companies. Their objectives are bound to differ in theory and practice from the old authorities. Private companies listed on the Stock Exchange will have to show a profit and provide shareholders with some sort of return on their investment. Sewer maintenance is strictly a non-profit making activity. The enormous amount of cash needed to replace crumbling sewers is unlikely to be found.

The Labour Party's spokesperson on the Water Industry, Ann Taylor, called for legal amendments to the Water Bill forcing the private companies to invest in the sewers, suggesting that a set of minimum standards be written into the Act. Environmental health officers proposed that a 'rat-baiting' clause be included in the Bill, giving guarantees similar in principle to the 'rural telephone boxes' clause incorporated when British Telecom was privatized. BT was forced to continue providing a rural telephone box service, despite the fact that it was regarded as a loss-making activity.

The Water Act, however, went through Parliament with no such amendments. The Director General of Water Services is the new regulator who is responsible for granting licences to the water companies. His powers include agreeing the levels of investment proposed by the new companies, but his guidelines for doing this remain vague. Individual water companies will draw up their own projected investment levels to the satisfaction of the Director General before the contracts are signed. The regulator, however, has a further duty to ensure the payment of dividends to shareholders. Many people are concerned that these two responsibilities will result in a conflict of interests with the provision of shareholders' dividends taking priority over necessary sewer investment programmes.

Clause 68 of the Act allows the Secretary of State to make regulations on standards of performance in connection with the provision of sewage services. It is essential for the proper maintenance of the sewer system that the Government takes strong action under this section, to ensure the new water companies undertake significant capital investment and sewer baiting programmes by insisting on high standards of performance. Clause 68 should be used for this purpose and the Government could, in appropriate cases, give grants to the new companies towards sewer maintenance programmes.

Chadwick's sewers have lasted in many cases for over 150 years, but they were not built to last forever, nor to take the large amount of sewage that we now produce, nor to withstand the heavy weight of modern traffic. The water companies will have to make a financial investment on the same scale as the Victorians if we want the sewers to last for another 150 years and keep the rats at bay.

Under the 1949 Pest Control legislation, local authorities are responsible for rodent control. As we have seen, if the surface infestations occur on private property, the local authorities have powers of entry, inspection and, if necessary, destruction of rats and the ability to seek compensation in the courts. Local authorities operate many public health-related services, including sewer-baiting in conjunction with the water companies, baiting other areas of public land, pest control services, rubbish collection and disposal and the maintenance of council property. Local government is undergoing an important change in direction, and many authorities have complained at what they see as the Government's obsession with making their services cost effective resulting in intolerable cut-backs in many areas of public health. It is no coincidence that the very councils worst affected by local government cuts to services are the ones which are suffering the most from substantial increases in rat infestation. Local authorities need to be adequately funded if they are to continue their role providing the community with basic public health services.

In the short term, a specific problem-related grant should be made available from central Government to local authorities to cope with the current rat crisis. The process of rodent control is costly and labour intensive – involving trained personnel, equipment, chemicals and vehicles. The grant should be allocated on the basis of local infestation rates and must be used specifically for rodent control, including baiting, regular maintenance work and ensuring the provision of adequate refuse collection and disposal.

Urgent action must also be taken over the national shortage of environmental health officers; local authorities are currently working with more than 500 vacancies. The staff shortages have had a serious effect on the maintenance of proper standards of health and rodent control. The Government must do more to encourage graduates into the field of environmental health and local government by creating special incentives and instituting a campaign of national recruitment.

On a local level, the councils themselves must recognize the need for more inter-departmental communication. There have been too many examples of housing, cleaning, engineering and pest control departments working against each other. The pest control department should be consulted at all levels within the authority. For example, with housing developments at the demolition stage to ensure that the drains are properly sealed and at the building stage to advise on rat-proofing. Housing departments should be talking to cleaning departments and constantly updating each other on local rat problems and potential areas of rat harbourage. By ensuring the pest control department is up to date in all aspects of rubbish build-up in its area, the risks of large scale rodent infestations will be greatly reduced.

In the long term, the wider issue of public health must be considered. Treating the present problem of increasing infestation by throwing one-off grants at it is only a short term solution. The Government should consider taking a wider view on rat baiting, sewer and drain maintenance, rubbish collection and disposal. Manchester's Chief of Environmental Health, Mile Eastwood, comments:

> If more money was available for long term prevention, authorities wouldn't have to make the kind of public health judgements

between what they ought to do, and what they can afford to do. If I want to fix a gate by a brook to stop children falling into the water where rats have been swimming and stop them potentially catching Weil's disease, I have to take the money out of my baiting programme, which is our long-term prevention against rats increasing. So, where does the money come from?

If the Government undertook a long term national programme of significant rat-related grants to local authorities over a ten year period, while at the same time embarking on a policy of substantial investment in the sewer system through the new water companies, the rat problem would virtually disappear.

Apart from cash injections to the local authorities and the water companies, there are other short term solutions which could help significantly reduce Britain's growing rat population.

New building practices should be urgently considered. Little thought has been given over the last decade to rat-proofing when demolishing old sites and constructing from scratch. Many rats gnaw their way into old and abandoned properties, or escape through sewers broken by diggers. Tighter working practices by demolition teams could ensure that their operations do not encourage rats. Attention to old drains, toilets, sinks and baths is important; they should be properly sealed. Preventing rats escaping from the sewers is an essential part of rodent control. Similarly, standard building procedures should include checking for old and disused sewers, and sealing broken drains where appropriate. All sites should be baited to extinction before developments are built.

Finally, architects and planners should involve rodent controllers when designing their buildings; a little preventative thought at the blueprint stage could go a long way to preventing rat infestations in the future. Where possible, harder, stronger materials should be used for drains. As we have seen, rats have no difficulty gnawing their way through plastic. Rodent controllers should attend council planning committee meetings adding their advice and, if necessary, instruction to potential applications.

In the countryside, the agriculture industry should increase its pest control budget. In many cases, farmers and landowners have stopped reporting rat infestations to local authorities altogether. Greater attention must be paid to methods of food storage and rat-proofing. Some farmers have been accused of the indiscriminate use of warfarin, contributing to the development of resistance in those rats. Local authorities should take on a stronger enforcement role in these cases. As previously mentioned, some local authorities want legislative changes made to the 1949 Prevention of Damage by Pests Act to include agricultural land under their jurisdiction and to give a more specific definition to the term 'substantial numbers' when referring to rats.

Owners of public land and public utilities such as London Regional Transport and British Rail should be made more aware of the dangers of rat infestations and their responsibilities towards rodent control. More money should be budgeted for baiting and other treatments. Environmental health officers are currently calling for the eventual prosecution of those who fail to provide adequate rodent control on their land. The IEHO survey revealed that many local authorities complained about British Rail's poor response to rodent control problems. British Waterways, London Regional Transport and other utilities would appear to be also cutting back on their operations. The Act has been with us for 40 years and it has the powers. It is time that EHOs rigorously enforced it if public utilities and landowners refuse to act on their own initiative.

An immediate public campaign should be initiated by the Government to educate the public about the dangers presented by Britain's increasing rat population. The public should be made aware, through television, radio, newspapers and magazine advertisements that the dropping of food and drink in the streets, the discarding of wrappers, hamburger cartons, bottles and cans, results directly in an increase in the rat population. The campaign should point out the threat to public health that rats pose, using vivid pictures of rats eating recognizable litter carelessly dropped by thoughtless members of the public, and highlighting the fact that half eaten take-away food is especially attractive to rats.

It should also bring to people's attention the importance of using dustbins. Families who regularly use more black bags than dustbins should buy extra dustbins, instead of allowing rats easy access to their waste. Again, striking and shocking footage should be used to illustrate the point, making people understand that it is the responsibility of us all to be more careful with our litter and rubbish and that by our actions, we can help prevent the increase in surface infestations of rats. Now is a good time to launch such a campaign, when public and politicians alike appear to be in agreement over environmentally sound issues; there is nothing greener than grass uncontaminated with rubbish and waste.

On the subject of public education, a more specific campaign should be launched to increase awareness of leptospirosis or Weil's disease. It should be aimed at both those who work regularly with or take part in leisure activities involving water and at the medical profession. As Dr Waitkins of the Leptospira Reference Unit in Hereford has indicated, we are only seeing a tenth of the real picture, which is worsening significantly each year. People are dying of the disease, often healthy people who could have been treated with a simple course of penicillin if the symptoms were recognized early.

The British Canoe Union issues its members with cards warning them of the dangers of Weil's disease, how to recognize it and what to do if they think they have it. It is an example other organizations should follow. Farmers, waterworkers, rodent control operatives, those taking part in water sports like open swimming, windsurfing, sailing, rowing and fishing should all carry cards and know the symptoms for which to look.

Doctors should be targeted for special attention. They should be aware of the possibility of Weil's disease amongst their patients, and advised of the existence of the Leptospira Reference Unit. The ELISA test should be used throughout Britain as it is in the rest of the world; its 3–7 day results could mean the difference between life and death. More money should be made available for research into leptospirosis and finding a vaccine; a direct Government grant for this purpose should be given to the PHSL Leptospira Reference Unit in Hereford.

But however effective a campaign promoting greater awareness of Weil's disease and pushing for more research funds might be, the basic problem remains, as Dr Waitkins confirms, the existence of rats in large numbers. Get rid of most of Britain's rats, she argues, and Weil's disease disappears too.

The current rat crisis can be effectively dealt with by implementing the suggestions in this chapter: short-term Government grants to the worst-affected areas of rat infestation; specific Government-funded programmes for sewer maintenance in the long term; co-operation between water companies and local authorities and the adoption of new national formulae for calculating rodent control allowances; effective campaigning on litter and leptospirosis; more resources from other public utilities for rodent control; better methods of building and planning and changes in agricultural practices. These policies can be put into immediate effect at a local level in close co-operation with the relevant local authorities.

What becomes increasingly obvious, however, is the need for a national co-ordinating agency with a specific brief for rodent control. The organization should take the form of a national office, created and funded by the Government for the specific purpose of monitoring and controlling the problem of rodent infestations in the UK. The role of the new body should be to co-ordinate all aspects of rodent prevention, from sewer baiting and maintenance to the administration and allocation of government funds.

Under the 1949 legislation, the Ministry for Agriculture, Fisheries and Food became the central co-ordinating body for rodent control in the UK. In 1980, as a result of the Local Government Planning and Land Act, many of the sections from the 1949 Act were repealed, and power was moved away from central Government to local government. It was said at the time that the local authorities were not keen on what they saw as 'central Government interference' in local matters. It was a short-sighted view, which has contributed to a national increase in the size of the rat population. Since no government body is monitoring the current situation, we have no official idea of the exact number of rat infestations in the UK, nor the size of the increase. Relying on private institutions like the IEHO and the AMA to commission surveys and collate figures should not be the Government's answer to a problem of national importance.

MAFF still retains some responsibility under the Act to provide advice and control measures on rats in Britain (for which it charges a fee), and to promote the need for rodent control through articles in the press and trade journals. MAFF is the obvious choice to create a new body to co-ordinate, collate and allocate resources.

One of the jobs of the new organization should be to collect and collate rat infestation information covering the whole of the country, allocating its resources to the areas which are the worst affected. It should be funded with two distinct budgets, one for programmes of short-term treatment and another for long-term prevention. The body should work with the local authorities and the new water companies. Once agreement is reached over the exact levels of local rat infestation on the surface and in the sewers, a national formula should be applied which would govern the financial allocations. Local authorities should therefore receive a grant from the water companies based on a uniform formula calculated by the new body. Increases in rat infestations would then lead to increases in funding. Special treatment grants taking into account local needs and resources should be paid by the new body to both local authorities and the water companies. This should include long term investment programmes covering sewer maintenance and other preventative activities, like extra money for rubbish collection and sewer building where appropriate.

The body should also act as a watchdog, monitoring those individuals, companies and public utilities which fail to treat rat infestations. It should inform the relevant authorities of the picture nationally and, in close association with them, embark on a policy of direct action recommending that the local authorities take appropriate measures through the courts. Faced with a national campaign of local legal action, British Rail, for example, would be forced to take its rat problems more seriously than it does at present.

The new MAFF body should also take on board the organization of the public education campaigns. This should include the immediate launching of a public health campaign, highlighting the dangers of rats and litter, and liaison with institutions like the Leptospira Reference Unit on campaigns covering Weil's disease.

In short, a new national co-ordinating body, created by resurrecting an organization already established under the 1949 Act, working closely within the local government framework, should be able to solve Britain's national rat problems on a local level. Its first target should be the immediate control of the current situation – short-term measures aimed at bringing down the size of the rat population – followed by long-term planning and co-ordination, effecting programmes of major investment and preventative action.

There is little that can be done to change Britain's weather; the recent mild winters look like being with us for a long time to come. What we can do, as individuals and as a country through our elected leaders, is to mitigate the effects of a mild climate on the rat problem by making a strong commitment to rodent control. We must put pressure on the Government to establish a national body responsible for rodent control with an adequate budget to deal with the current increase in infestations. Each of us must be more responsible for our own litter and rubbish.

Rats represent a real threat to our public health and safety. If we are to learn anything from history, it is that rats should be exterminated quickly and efficiently, before their numbers grow out of control. Failure to act now means living with a potential time bomb in our midst as we suffer from an ever-increasing number of rats – and the risk of a new plague.

Select Bibliography

BOOKS

Martin Hart *Rats* (Allison and Busby, London, 1973)

Mark Hovell *Rats and How to Destroy Them* (John Bale and Sons Danielsson Ltd, London, 1924)

William H. McNeill *Plagues and People* (Basil Blackwell, Oxford, 1977)

Henry Mayhew *Mayhew's London (1861)* (Bracken Books, London, 1984)

A. P. Meehan *Rats & Mice* (Rentokil, Surrey, 1984)

Brian Plummer *Tales of a Rat-Hunting Man* (Robin Clark Ltd, Milton Keynes, 1978)

Graham Twigg *The Black Death* (Batsford Academic and Educational, London, 1984)

Graham Twigg *The Brown Rat* (David and Charles, 1975)

Philip Zeigler *The Black Death* (Collins, London, 1969)

Hans Zinsser *Rats, Lice and History* (Little, Brown and Co., USA, 1934)

REPORTS

'Cases of Leptospirosis in Britain 1978–1988', Communicable Disease Surveillance Centre

'Rats on the Rise', Association of Metropolitan Authorities Survey, February 1989

'Rats – A Survey of Incidence of Rats in England and Wales,
    1988–1989'
'Report by Institution of Environmental Health Officers',
    August 1989

ARTICLES

'Rats are on the Move', *Essex Countryside*, June 1978
'Rattus Rattus: the Introduction of the Black Rat into
    Britain', James Rackham, *Antiquity*, 1979
'New Evidence of Black Rat in Roman Britain', P. Armitage,
    B. West, K. Steadman, Museum of London, 1983
'The Legend of the Pied Piper', Christopher Nicholson, *The
    Listener*, 19 July 1984
'Royal Ratcatcher's Sash', *ServiceScene*, Rentokil, February
    1986
'The Long-tailed Fellows', Peter Bateman, 1986
'Plague of Rats Escape Sewers', *The Observer*, 18 September
    1988
'Weil's Disease', Dr John Whitehead, BCU Medical
    Advisory Panel, 1988
'Keep Hertsmere Tidy', *Hertsmere Connection*, Summer 1989
'Sewer Rat Warning', *The Guardian*, 16 February 1989
'Rodent Control in Public Health', *International Pest Control*,
    February 1989
'When is a Pest a Pest?', *New Scientist*, 4 March 1989
'Rat Disease Peril in the Park Water', *Evening Standard*, 11
    August 1989

Index